T0092856

SAS® Coding Primer and Reference Guide

SAS® Coding Primer and Reference Guide

Connie Kelly Tang

CRC Press is an imprint of the
Taylor & Francis Group, an **informa** business
AN AUERBACH BOOK

First edition published 2020
by CRC Press
6000 Broken Sound Parkway NW, Suite 300, Boca Raton, FL 33487-2742

and by CRC Press
2 Park Square, Milton Park, Abingdon, Oxon, OX14 4RN

CRC Press is an imprint of Taylor & Francis Group, LLC

ISBN: 978-0-367-50794-7 (pbk)
ISBN: 978-0-367-53705-0 (hbk)
ISBN: 978-1-003-05129-9 (ebk)

Typeset in Minion
by codeMantra

Contents

Preface

I have been using SAS since my undergraduate days, and I am a SAS-certified base programmer and practitioner. I am not, however, a daily SAS user. When I need further explanation or clarification on SAS coding, I turn to the web and the SAS community. While I have never been disappointed with the voluminous material out there on the web, I have concluded that a systematic introductory-level guide with different scenarios for similar nondaily users would be highly valuable. Additionally, applied statistical concepts would be immensely helpful to practitioners like me who are not majoring in statistics, but, nevertheless, use statistics for research and development.

This book is my attempt to deliver this introductory-level guide. All material is based chiefly on notes I have taken and practical sample data I have worked on over the years.

The material covered in this book will serve not only as a textbook for those who want to learn SAS, but also as a reference guide. Users can get a quick-start on their effort by customizing the relevant codes I have provided for various tasks in this book.

Connie Kelly Tang

READ ME FIRST

To best take advantage of the material in this book, please be aware of the following five key aspects:

a. SAS commands or SAS keywords used in all my examples are highlighted with a gray background color. You can't change or modify these grayed commands or keywords. They must be spelled out exactly as they are presented. The gray background color helps you to identify what you can or can't change or modify. You can rename and respell words, phrases, notes, and names that aren't grayed out.
b. Make sure you understand the data used in an example. Some examples have their entire data displayed. Others have only small portions displayed as the focus is to illustrate data formats and structures.
c. Follow the right-side code explanations if you get stuck on how a set of statements works. I try to explain every statement in a SAS code file, even it means repetition. My goal is to make the information both easy to follow and easily accessible, so you don't have to flip back and forth through this book.

d. For Chapter 7, start with the chart/figure first. From there, go over the data and understand how the data are visualized in the chart/figure. Study the SAS codes generating the chart/figure last.
e. The vast majority of examples have their corresponding SAS codes. If you want to run such codes by copying and pasting them to your SAS editor, check out the quotation marks “ ” or the apostrophes ‘ ’ to ensure that they are still proper. In my own testing among different computers, occasionally, the copying and pasting actions lead to incorrect quotation mark and apostrophes.

Author

Connie Kelly Tang is a graduate of New York University (NYU), Abu Dhabi, with a BA in Economics. She also holds a master's degree in Applied Economics from the University of Maryland. She has learned and utilized SAS in her private consulting after graduating from NYU in 2014. Since then, she has provided technical assistance to a wide range of activities and programs related to data, including her near 3-year services as an on-site consultant to the Office of the Secretary, the U.S. Department of Transportation. Currently, Connie Tang is the assistant director for research and outreach at the Maryland Transportation Institute, University of Maryland, supporting and leading a broad range of research and development activities.

1

Basic Components of SAS

1.1 SAS FILES

You will encounter four different basic file types when using SAS for data analysis. These are as follows:

1. SAS code file
2. SAS data input file
3. SAS log file
4. SAS output file.

- *Code file*
- *Data file*
- *Log file*
- *Output file.*

Normally, in SAS data analysis, you obtain your data first. After that, your task is to understand your data's structure, its variables, and the questions you are trying to answer. From there, you write your SAS codes and perform the analysis. Lastly, you will try to understand your results, which will help you to develop new code to rerun your data from a different perspective. The SAS log file and output file are automatically generated once you run your SAS code file.

1.1.1 SAS Code File

Your SAS code file contains all your codes, statements, and procedures. It may even include the actual data that you have for a project. The SAS code file is a text file and has the **.sas** file extension.

You will most likely use the **SAS Editor** to write your SAS code. However, you can write and edit SAS codes in any text editor software (e.g., Notepad, WordPad).

You can name your SAS code file by any combination of characters, numbers, spaces, and symbols. Unlike the SAS naming conventions for variables and others, there is no length, character or symbol, uppercase or lowercase limitations

Code File Editors

- *SAS Editor*
- *Notepad*
- *WordPad.*

when it comes to naming a SAS code file. However, you should follow the good file naming practices discussed in Section 1.5.

Naming Code Files

- *Anything goes but follows good practices.*

Example: SAS code file names

Project air OD UMD version 1.sas
UMD 2019_December project air OD version 1.sas
Dec19_UMD_air_OD_v1.sas
UMD_p78_v3.sas

1.1.2 Data Input File

SAS can read a wide variety of data in a wide range of formats. SAS's own native data file has the .sas-7bdat extension.

SAS Data File Extension

- *.sas7bdat*

You should always ensure you fully understand your data and data structure, and know what variables (columns) you have, variable types (character vs. numeric vs. date, etc.), and how they are related to each other. Your data understanding is not only crucial for programming your SAS code effectively but also vital to make sure you utilize appropriate statistical methods in your analysis.

1.1.3 SAS Log File

A SAS log file is automatically generated when your SAS code file is executed. By default, it is an HTML file.

Log file

- *Autogenerated.*
- *Review it always.*

The SAS log file contains essential information on how codes, statements, commands, and data are executed line by line. You should always review your log file carefully, even if your run produces outputs in the output window.

1.1.4 SAS Output File

The product of executing your SAS code file is the output file. You can have more than one output file depending on what you instruct SAS to do in your SAS code file. SAS's default output files are in HTML format. However, SAS can deliver outputs, results, charts, figures, and maps in a host of other formats.

Outputs

- *Could be more than 1 file.*

1.2 ORGANIZING A BASIC SAS CODE FILE

Simply put, you can compare using SAS to using an automated mortgage or car loan amortization spreadsheet where you just put in your loan amount, the length of the loan, and the interest rate. The amortization spreadsheet will automatically generate a monthly payment schedule, including monthly payment, principal pay, interest pay, and cumulative principle related to the loan. SAS will do the same for you. Powerful and sophisticated computational procedures associated with SAS have already been programmed internally. You just need to know what information you have, what you want, and the procedure you need. By listing these parts together in a specific order (syntax and grammar), SAS automatically analyzes the data and delivers results.

1.2.1 Syntax

Your SAS code file should be written in chunks of statements. There are three major building chunks when it comes to putting your SAS code file together. The first chunk starts with the "**DATA** your_data_name;" statement. This step is commonly referred to as the "DATA" step." This step is usually where data manipulation and variable manipulation are performed. The DATA step must end with the "**Run**;" statement. Before we move ahead to the other two chunks, keep in mind that all chunks end with "Run;" statement. See examples below.

```
DATA   new_test_score;
.
Y2=x1+x2
.
Run;
```

DATA creates a new dataset called new_test_score.
The new dataset is in the temporary WORK folder. We know it because there is no specific Library (folder) specified. The default is the WORK library.
A new variable Y2 is created by adding variable x1 to variable x2.
Run is a SAS command. It kicks off the DATA step here.

The second chunk type is the **PROC** step. This step includes statements such as PRINT, MEANS, and ANOVA. As stated earlier, all statements end with the "**RUN;**" statement.

```
PROC MEANS;
.
VAR y1;
.
RUN;
```

PROC MEANS is a SAS command. It produces a host of mean-related statistics for variable y1 as specified by the VAR command.
RUN kicks off the PROC MEANS procedure here.

The third chunk is something I call the miscellaneous. Statements found under this chunk are often global in nature. When such statements are executed, all other blocks are affected.

These categories include titles, footnotes, paper size, line size, format, and output delivery. See examples below.

```
options linesize= 80
pagesize= 60;
portrait **;
```

options is a SAS command. Specifications listed are executed for all SAS codes and are valid until the options are revoked.

linesize=80 means a line can have a maximum of 80 characters (columns).

pagesize=60 means a page can have a maximum of 60 rows.

portrait directs the output in a portrait direction vs. the landscape orientation.

1.2.2 Basic SAS Grammar

Semicolons (;)

Every statement must end with a semicolon (;).

Asterisk (*) and Forward Slash Key (/)

When you write your SAS code, it is helpful (and recommended) that you include comments. Comments are lines in your code file that SAS ignores and does not execute. Comments help you or people in reviewing or using your code to better understand your program. Because comments are ignored by SAS, it will not face the same syntax restrictions.

Comments

- *For your own reference*
- *Ignored by SAS*
- *Starts with * and ends with ;*
- *Or starts with /* and ends with */*

There are two ways of adding a comment to your SAS code.

In the first method, an asterisk * marks the start of your comment. After you place the * down, you can then begin writing your comment. When you are done writing your comment, just close your comment off with the semicolon (;).

The second method is the exact same as the first, except you use a slash asterisk (/*) to start the comment and */ to indicate the end of the comment. The method is very convenient for long lines of text that you do not want SAS to execute.

See the examples illustrated below:

*the above statement is to compute the subtotal length of all highways based on variable kmL;

/*   the above statement is to compute the subtotal of all roadway length data based on variable kmL. Pay attention to the kmL, which is a derived variable from step 981.

And the third-party codes should not be used.   */

Cases of Alphabets (Letters)

SAS does not differentiate uppercase or lowercase letters used in variable names, file names, and statements. You may, however, elect to use a combination of uppercase and lowercase letters to improve human readability of a variable name or statements.

Cases of Letters

- *SAS does not differentiate uppercase and lowercase letters.*
- *They are treated as the same.*

Blank Spaces

SAS ignores extra blank rows and blank spaces between variables and between statements.

Unlike many other software languages, when it comes to listing variables, SAS uses spaces to separate variables, not the commas.

> **Blank Spaces when Listing Variables and between Rows**
>
> - *SAS treats any columns of blanks as 1 blank (counted as 1 column).*

You can list variables and others (e.g., observations) in multiple rows as long as the row only ends with a single semicolon (;).

Quotation Marks

You should use quotation marks ("string") or apostrophes ("text") to enclose a string (text) observation.

@ Symbol

The single trailing @ holds on the observation for a given variable until statements following the symbols are executed.

@n Symbol

The trailing @n symbol directs SAS to jump to the nth columns to read the variable.

@@ Symbol

The doubling @ @ symbol directs SAS to continue to read data in sequential order.

% Symbol

SAS Macros always use the % symbol, but not all % are Macro-related.

sas7bdat Extension

The default SAS data has the .sas.7bdat extension.

1.3 SAS DATA DIRECTORY AND STORAGE

In the Windows environment, a file is stored in a folder by following a specific file directory. For example, you can store an Excel file called "**score.xls**" in the folder of "**newdata**" under the directory structure of "**c:\data\newdata**."

In SAS, the directory structure, as illustrated by **c:\data\newdata**, is called a library. You can give the library a name (e.g., *mydata*) using the **LIBNAME statement** as illustrated below.

```
LIBNAME mydata  'drive:/
directory' ;
```

LIBNAME defines a mydata library representing the directory or the path of "drive/directory."

This is particularly helpful when you have a multilevel long directory structure (e.g., c:\umd_projects\p2021\phase1\stage2). Once a directory is defined by the **LIBNAME** statement, you can refer to the library name instead of the full directory.

Example

```
LIBNAME     NewD1
'c:\data\MyNewData\test_score' ;
```

LIBNAME defines NewD1 as the library name representing the file directory of c:\data\MyNewData\test_score.

Here, the directory of c:\data\MyNewData\test_score is defined with a library name of NewD1.

In the future, the directory can be referenced as NewD1 as opposed to c:\data\MyNewData\test_score.

A two-layer naming convention is used to name a SAS data file. The first layer is the library name. The second layer is the actual file name. They are joined together by the period (.).

SAS Data File Naming Convention

- *Two layers*
- *First layer: library name*
- *Second layer: filename*
- *The first layer is joined to the second layer by the symbol of . period.*
- *NewD1 in NewD1.Class36 is the first layer – library – and Class36 is the second layer – the actual file name.*
- *When no library is specified, the file resides in the default WORK library.*

Example

You have a file named **Class6score.sas7bdat** stored at c:**\data\MyNewData\test_score** folder. Instead of referring to the whole folder directory, you can simply use **NewD1.Class36score** instead.

When a file is only specified or listed by the second layer (the actual file name, e.g., Class36score) without referencing a library, the file is assumed to be located in the default library of **WORK** (we will cover **WORK** in Chapter 2). Files resided in the **WORK** library are temporary, meaning that when SAS is terminated, these files will be deleted automatically.

1.4 TYPES OF SAS VARIABLES, INFORMAT AND FORMAT

You will encounter three types of variables in SAS.

- Numerical (e.g., 0.125, –12.3, 45.7%)
- Character (text, string) (e.g., dog, all89)
- Date/Time (e.g., 12/3/2025, 18:36:45).

When SAS reads data in from a file or from its embedded date, you can specify variable types via **INFORMAT** statements. **INFORMAT** statements define variable types, variable width (# of columns), formats used in coding date and time, and other aspects of a variable. It guides SAS to read the data correctly. More coverage on SAS variables is provided in Section 3.2.

Once a dataset is read correctly, SAS delivers and displays the output. You can customize the appearance of observations and analysis results in the output through the statement of **FORMAT**. For example, you can use the **FORMAT** statement to control how many decimal points should be displayed for a numerical variable (e.g., 3.61289 vs. 3.61), the way a date should be displayed (e.g., 12/21/2028 vs. December 21, 2028), and the displayed width of a variable (e.g., ABC vs. AB). Be aware that **FORMAT** statements do not change the actual data. They simply change how the data are visually presented.

- **INFORMAT:** controls how data are read into SAS. It affects actual data.
- **FORMAT:** controls how data are displayed or presented in output (appearance only). It does not change the actual data.

INFORMAT

- *How SAS reads in data.*
- *Affects actual data.*

FORMAT

- *How SAS displays data.*
- *Doesn't affect actual data.*

1.5 GOOD PRACTICES FOR NAMING SAS FILES AND VARIABLES

1.5.1 Naming SAS Code Files

How you name and store your file is critical to how quickly you can locate it and know its contents.

The names you choose should be descriptive, consistent, and human-readable. You should store your files using a clear directory structure, such as by project name, date, client, and other unique identifiers.

Tips for naming files:

- Keep your file name short.
- Do use descriptive human-readable words, phrases, abbreviations, letters, numbers, and other symbols.
- Do use underscores to delimit words.
- Do use capital letters to delimit words or abbreviations.
- Do not use spaces to delimit words. Do not use the following for names as they are specifically prohibited in SAS:

NULL
DATA
LAST
_N_
ERROR

Naming SAS Variable

- *Can be as long as 32 characters (columns).*
- *Follow good practices.*

1.5.2 Naming SAS Data Variables

Data in SAS are organized by variables (column names). You should generally follow the same principle for naming variables as for naming files. Specifically, you should follow the specifications listed below.

- Variable name length – up to 32 characters in length
- The first character – must be either an English letter or an underscore
- No blank, no space, and no special characters other than the underscore _
- Letters are not case sensitive in SAS. While SAS treats uppercase and lowercase letters the same, you can still use uppercase and lowercase to help the readability of a variable. For example, FirstScore is more readable to users than firstscore.
- Do not use _N_, _ERROR_, _NUMERIC_, _CHARACTER_, and _ALL_ for your variable names.

2

SAS in Windows Environment

2.1 RUNNING SAS IN WINDOWS ENVIRONMENT

When you start SAS in the Windows environment, you will see the following layout consisting of four different panels.

- Explorer
- Log – Untitled
- Editor – Untitled1
- Output (this will appear below the Editor panel).

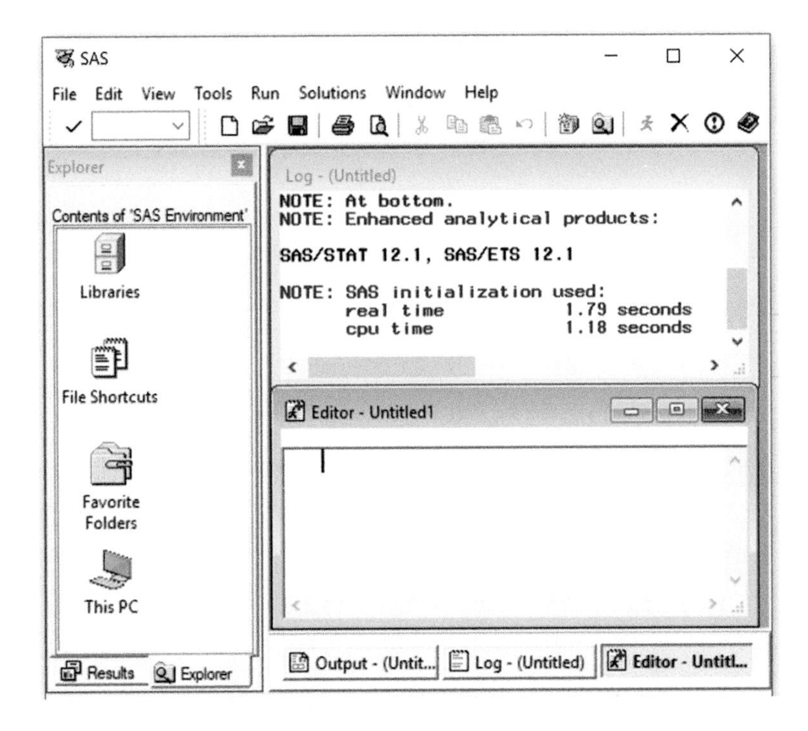

The top bar follows the standard Window dropdown menu format. When you click on it, submenus appear, from which you can make further selections.

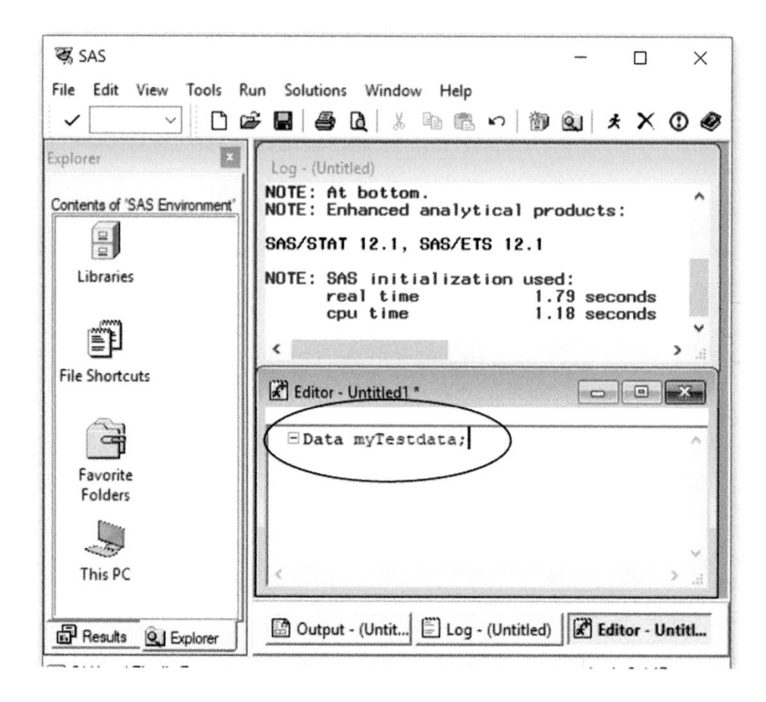

When creating your SAS code, you should type all commands and statements in the Editor panel.

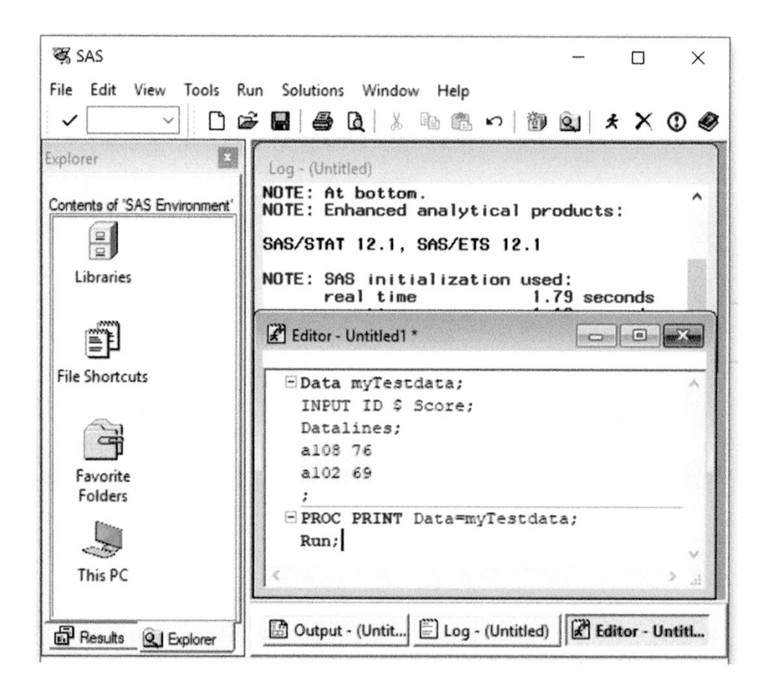

When you need to save your SAS code, click the Save As (if this is the first time you are saving the file or you want to save your file with a different name) button located with the submenu of the File menu.

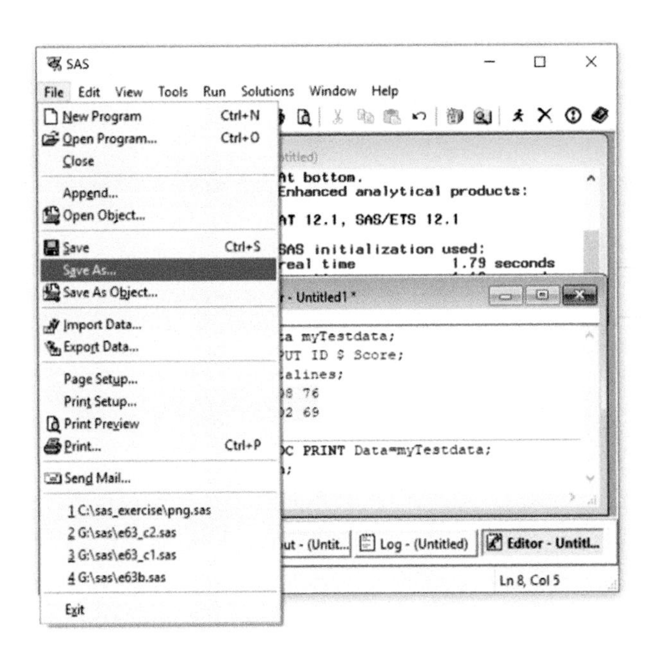

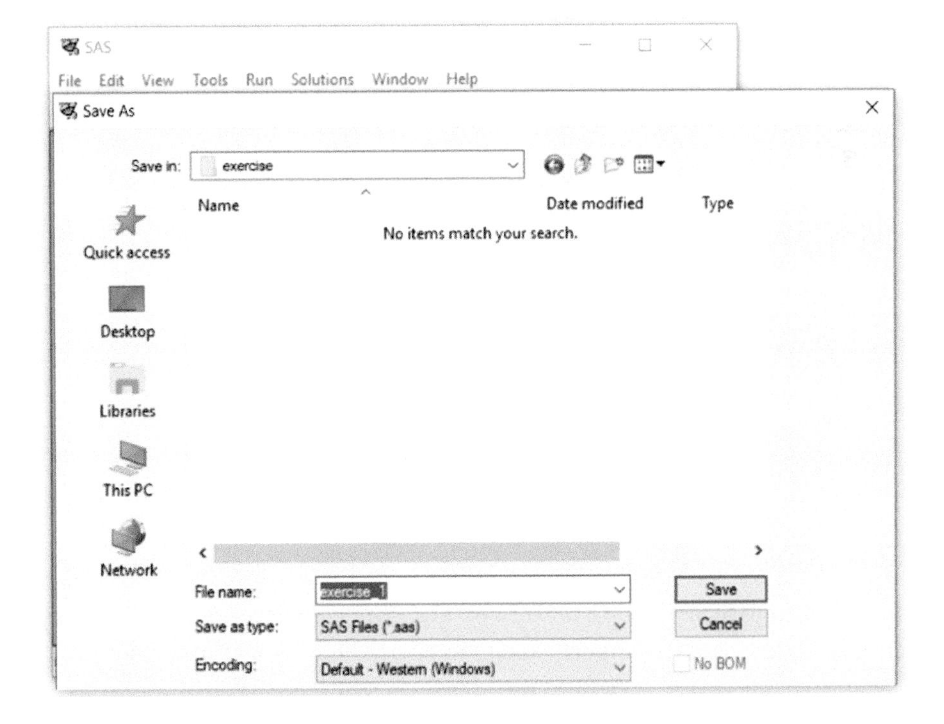

After your SAS code file is saved, you can **"Run"** your code file by pressing the run button.

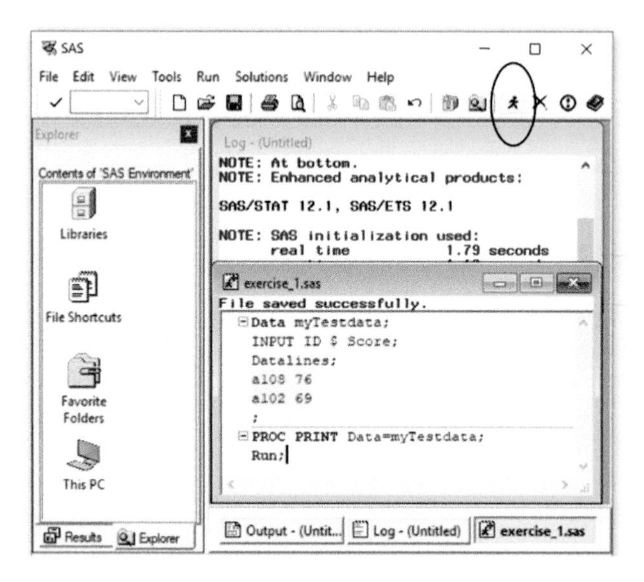

A new screen will pop up after your SAS code is executed. Results are displayed on the screen, as illustrated below.

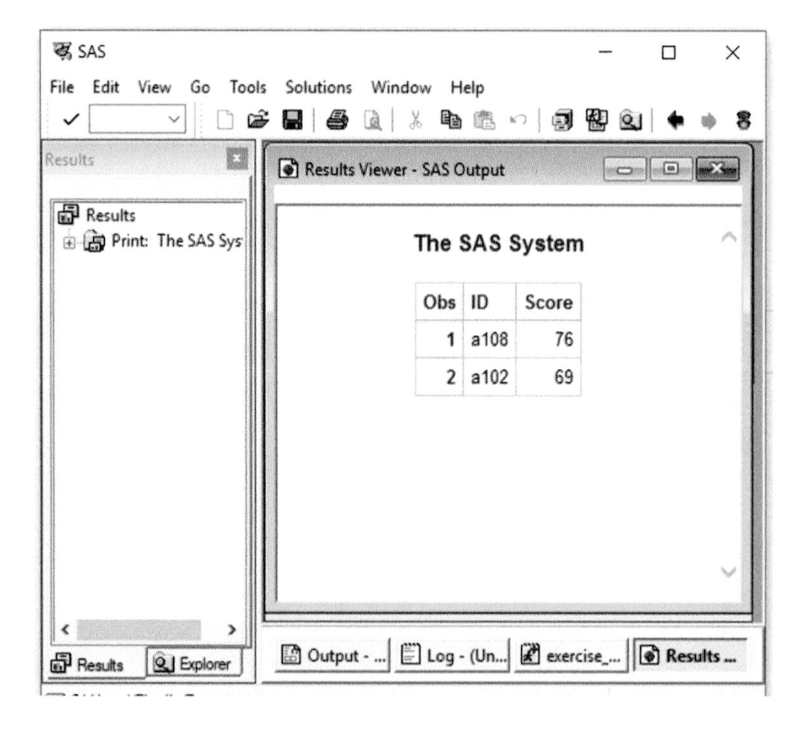

You will also automatically obtain a SAS Log file. This file will be generated regardless of whether your code has been properly executed or not. Always examine your SAS Log file to ensure all statements and data steps have been carried out correctly.

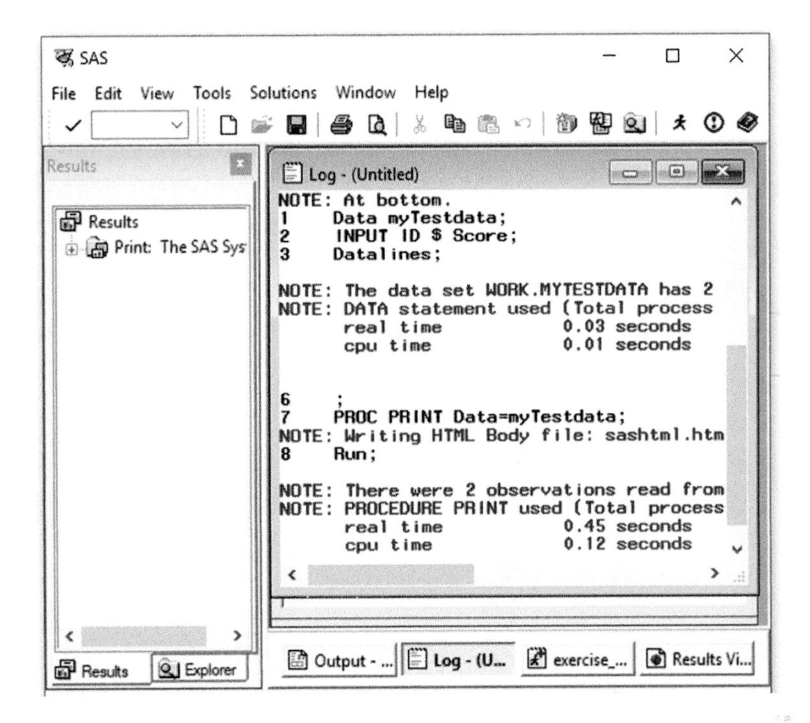

Remembering the Default Library – also known as the WORK library.

By default, all SAS-generated data are stored in the WORK library. When you close out of SAS, any data stored in the WORK library disappears. In the example below, mytestdata.7bsasdat generated in the process is in the WORK library.

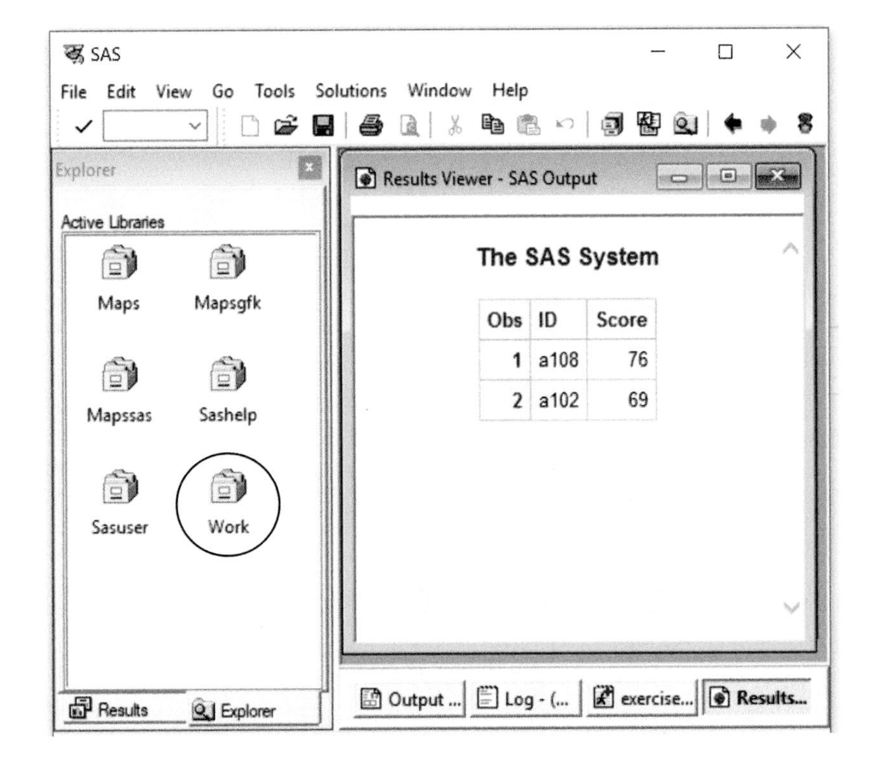

The WORK library (folder) is one of 6 default libraries.
- *It is SAS's* ***default*** *data storage library.*
- ***Mytestdata*** *in the* ***WORK*** *library is produced by the code* ***DATA mytestdata****. It has the .sas7bdat extension.*
- *You can use the* ***Explorer*** *button to navigate among all folders/libraries.*

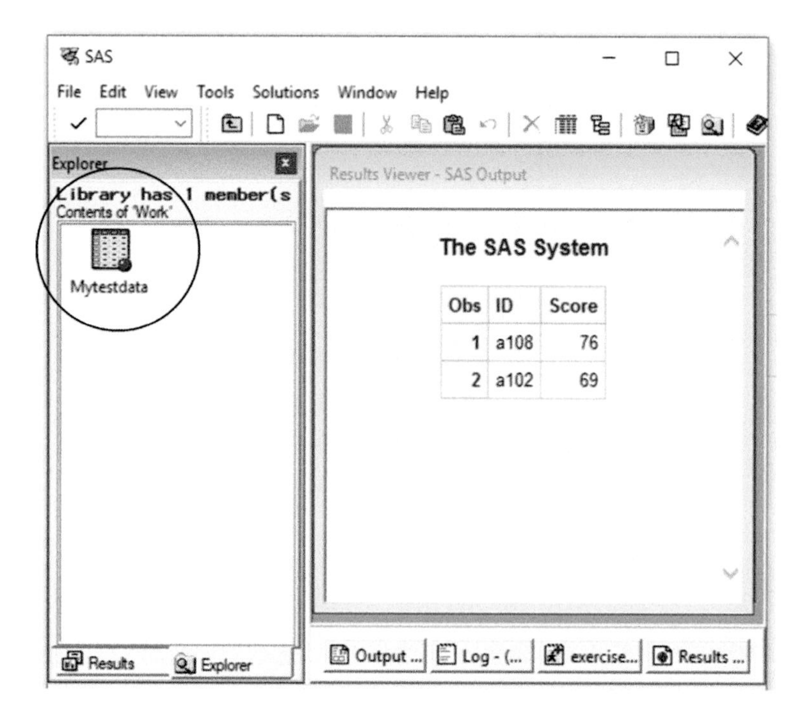

2.2 UNDERSTANDING SAS DATA EXISTENCE, STORAGE, AND DIRECTORY STRUCTURE

When a dataset is read or imported into SAS, it can be further stored in two different forms: temporary and permanent.

Remember – anytime you create a SAS data in the default library WORK – the dataset is temporary.

Examples

```
DATA mydata;
DATA WORK.mydata;
```

"DATA mydata;" creates a SAS dataset named mydata and stores it in the default WORK library.

"DATA WORK.mydata;" does the same thing. When a file is named with only one layer (mydata), it defaults to the WORK library.

The mydata data file created above is temporary. The first statement only uses a one-layer name: mydata. By not specifying a library, it defaults the location of your file to the **WORK** library. The second statement uses the standard two-layer naming convention with **WORK** being the first part and **mydata** the second part.

To create a permanent SAS data file, you will need to create a separate library – a unique library, as shown below.

```
LIBNAME     myown    'c:\data\MyNewData\my_score_data\year2019' ;
DATA myown.mydata;
```

Create a Permanent SAS dataset:
LIBNAME defines the directory structure of c:\data\MyNewData\my_score_data\year2019 as a library named myown.
"DATA myown.mydata;" creates a permanent SAS file named mydata.sas7bdat and stores it in the myown library, which is the folder of c:\data\MyNewData\my_score_data\year2019.

3

Feeding Data to SAS

3.1 PREPARING DATA FOR ANALYSIS

Before you can do any data analysis with SAS, you must first organize your data. As with any other relational database system, SAS data are made up of rows and columns. In SAS data, a row is called an observation, and a set of columns is called a variable (a single variable may occupy several columns). Table 3.1 illustrates an example of SAS data layout.

TABLE 3.1
Rows (18 Observations) and Columns (3 Variables)

LocationID	Date_Time	Speed_MPH
SR7901	8/1/14 12:00 AM	66
SR7901	8/1/14 12:30 AM	65
SR7901	8/1/14 1:00 AM	66
SR7901	8/1/14 1:30 AM	67
SR7901	8/1/14 2:00 AM	65
SR7901	8/1/14 2:30 AM	66
SR7901	8/2/14 12:00 AM	65
SR7901	8/2/14 12:30 AM	67
SR7901	8/2/14 1:00 AM	66
SR7901	8/2/14 1:30 AM	64
SR7901	8/2/14 2:00 AM	65
SR7901	8/2/14 2:30 AM	65
SR7901	8/3/14 12:00 AM	64
SR7901	8/3/14 12:30 AM	61
SR7901	8/3/14 1:00 AM	61
SR7901	8/3/14 1:30 AM	59
SR7901	8/3/14 2:00 AM	59
SR7901	8/3/14 2:30 AM	60

Observations – rows
Variables – columns

3.2 TYPES OF VARIABLES USED BY SAS

You will encounter three different variable types when using SAS:

- Numerical (e.g., 0.876, –12.13)
- Character/string (e.g., truck, abc12)
- Date/Time (e.g., 12/04/2020. 23:16:45).

Variable Types

- *Numerical*
- *Character*
- *Date/Time.*

When feeding data to SAS, you need to define variable types through SAS's **INFORMAT** statement. **INFORMAT** statements tell SAS how to read data and are commonly used in conjunction with **INPUT** statements.

- Character INFORMAT statement: var **$ w.**
- Numeric INFORMAT statement: var **w.d**
- Date/Time INFORMAT statement: var INFORMAT**w.**

  var – indicates the variable name
  $ – indicates a character variable
  w – indicates the width (number of columns) of a variable
  . – indicates to SAS that what listed is an INFORMAT specification
  d – used only for numeric variables specifying decimal points.

For numeric variables, **w** = (columns covering all whole digits) + (one column for the decimal point ".") + (columns used for digits for decimals). The **w** should always be larger than **d** or equal to d.

3.2.1 Character Variables

To specify a variable as a character type, you list a **$**, **w**, and (.) following the variable name. The **$** sign tells SAS that the proceeding variable is a character type. The **w.** tells SAS the maximum number of characters (length) a variable can have.

If you do not specify the **w.**, the default width is 8 columns. Observations with more than 8 characters will be truncated to 8 characters.

Character variable INFORMAT: $w.

You can also use the **:$w.** to let SAS read an observation up to the **w** columns. For example, **:$10.** indicates that the observation could be at most **10** columns wide, but not necessarily **10** columns wide. Some of the observations may be **6** columns. Others may be **3** columns, etc. Please notice the usage of the **:** mark with the above **INFORMAT.**

Examples

```
INPUT State_name $2. ;
```

State_name is a character variable indicated by the $ INFORMAT, and its observations have a width of 2 columns dictated by the 2. INFORMAT specification.

The above statement tells SAS that the variable **State_name** is a character type, and it has a width of 2 characters. The space separating the variable name **State_name** and the **$2.** is irrelevant.

```
INPUT       State_name   $2. ;
```

and

```
INPUT   State_name      $2. ;
```

and

```
INPUT    State_name      $  2. ;
```

are the same.

The empty spaces between the $ and the 2. do not affect the validity of the INFORMAT statement.

Reading Character Data Examples

Example a: Importance of specifying variable types

If a character variable is not specified as a character variable through the **INFORMAT $** symbol, the data will not be read. See the illustration below.

SAS Codes (E321a)

```
DATA testDefault;
INPUT Name ;
DATALINES;
Tom
Mary
Peter
Camila
Benjamin
Elizabeth
Clementine
;
PROC PRINT
DATA=testDefault;
RUN;
```

DATA command creates a new dataset with the name testDefault.sas7bdat and stores it in the default WORK library.

INPUT lists only one variable called Name in this example.

The variable Name as specified defaults to be the numerical type.

DATALINES tells SAS that actual observations (data) are next.

Actual data: Observations are listed and ended with the ; symbol.

Obviously, these observations are not numerical.

PROC PRINT is a SAS command that prints out the result on the screen.

The DATA to be printed is the newly created testDefault.sas7bdat in the default WORK library.

We know the testDefault dataset is in the default WORK library because only one-layer name is used.

Result:

The SAS System

Obs	Name
1	.
2	.
3	.
4	.
5	.
6	.
7	.

The printed result shows that all the observations for the variable Name are blank marked by the . symbol.

- *Data can't be read because the INPUT did not specify Name as a Character variable.*
- *Consequently, the testDefault.sas7bdat is empty for the Name field.*

Example b: Length of variable data and default length specification

The default length of a character variable is **8** columns. If your character data has more than **8** characters, make sure you specify the correct length **(w.)**. See all illustrations below.

SAS Codes (E321b)

```
DATA testDefault;
INPUT Name $ ;
DATALINES;
Tom
Mary
Peter
Camila
Benjamin
Elizabeth
Clementine
;
PROC PRINT
DATA=testDefault;
RUN;
```

This time the INPUT statement through the INFORMAT specifies that the Name is a character variable by the $ symbol.

Now, Name has the correct data type, and observations are read correctly.

However, observations (rows) 6 and 7 are truncated to 8-character width. Elizabet and Clementi do not have the full spelling.

This is because the "INPUT Name $;" did not specify length for variable Name. The default maximum 8-character length is used to read in data.

Result:

The SAS System

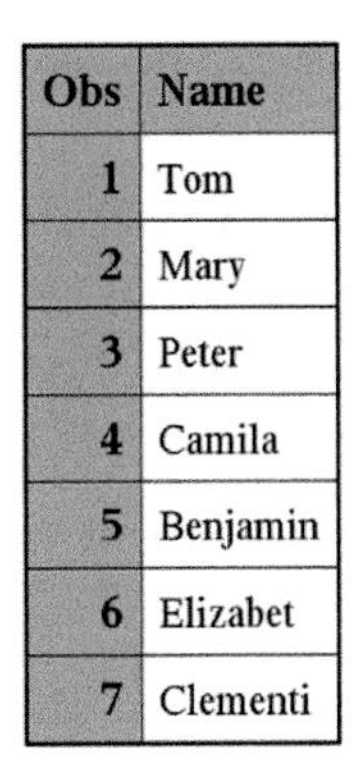

Obs	Name
1	Tom
2	Mary
3	Peter
4	Camila
5	Benjamin
6	Elizabet
7	Clementi

Example c: Length of variable data and customized specification

SAS Codes (E321c)

```
DATA testDefault;
INPUT Name $10. ;
DATALINES;
Tom
Mary
Peter
Camila
Benjamin
Elizabeth
Clementine
;
PROC PRINT DATA=testDefault;
RUN;
```

The "INPUT Name $10." specifies that the maximum width for any observations for the variable Name is 10-character wide, which is the same as Clementine.

Result:

The SAS System

Obs	Name
1	Tom
2	Mary
3	Peter
4	Camila
5	Benjamin
6	Elizabeth
7	Clementine

Example d: Specifying a length with an alternative customized specification

You can also use the LENGTH statement to define a character variable length, as illustrated below.

SAS Codes (E321d)

```
DATA testDefault;
LENGTH name $7.;
INPUT Name $ ;
DATALINES;
Tom
Mary
Peter
Camila
Benjamin
Elizabeth
Clementine
;
PROC PRINT DATA=testDefault;
RUN;
```

"LENGTH name $7.;" is another example of INFORMAT specification. It declares that observations with variable name is 7-column maximum.

Only 7 columns of observations associated with the variable name are read and printed out.

Result:

The SAS System

Obs	Name
1	Tom
2	Mary
3	Peter
4	Camila
5	Benjami
6	Elizabe
7	Clement

3.2.2 Numeric Data

The **w.d** format is the most often used **INFORMAT** statement for a numeric variable.

Here the **w.** represents the width of the input field, including the decimal symbol and all the decimal digits. The **d** specifies (a) the number of decimal digits for observations possessing decimal points and (b) the power of 10 by which to divide the value if the observation is an integer.

For example, **568** is an integer with an **INFORMATw.d** of **6.1**. SAS will read the **568** as **568**/10^1, which is 56.8. If the **w.d** specification is **6.2**, then **568** will be read as **568**/10^2, which is 5.68. In SAS, 568 is different from 568.0. 568 is an integer. And 568.0 is a real number with a decimal digit.

While working with numeric variables, you should specify a width long enough to hold your widest data value. This will prevent your value from being cut off. You should check out all examples below to fully understand the impact of the **w.d** INFORMAT specification.

INFORMAT for standard numeric data

- *w.d*
- *w is the maximum width (columns) of an observation*
- *d is the decimal point for non-integer data*
- *For integers, d is the power of 10 to be used as a divisor for final value determination*
- *w> = d.*

Example:

Examples below are designed to show the effects of **INFORMAT** on how SAS reads data. I used the **FORMAT 10.6** for the variable **td** to provide enough column width in order to show the **INFORMAT** effect. The **FORMAT 10.6** tells SAS to display the result of what being read by SAS through the **INFORMAT** specification with **6** decimal digits. This **FORMAT** specification still leaves **3** columns for the whole number displaying if needed (6 + 1 + 3 = 10).

Examples used here are to illustrate INFORMAT effect.
- *The only difference among examples is the INFORMAT.*
- *Pay close attention to the result.*

Case 1: INFORMAT for the Variable td is 8.4

SAS Codes (E322c1)

```
DATA test1;
INPUT td 8.4;
DATALINES;
12
123.0
123.1
123.12
123.123
123.1234
0.00005
0.0005
0.005
0.05
0.5
;
RUN;
PROC PRINT
DATA=test1;
FORMAT td 10.6;
RUN;
```

DATA creates a SAS dataset called test1 in the default WORK library.

INPUT lists the variable td, and its INFORMAT is 8.4 (8 columns in total with 4 decimal points).

DATALINES declares that actual data is embedded in and is next.

12 is an integer.

The rest observations for the variable td are real numbers with different decimal digits.

"RUN;" kicks off the DATA step.

"PROC PRINT DATA = test1" let SAS print out the newly created test1 file stored in the default WORK library.

"FORMAT td 10.6;" instructs SAS that printed (displayed) observations follow the format of 6 decimal points and with a maximum width of 10 columns.

Results show that all observations have 6 decimal points because of the FORMAT instruction.

Pay attention to how SAS reads in all data.
- *Given all observations conform to the INFORMAT of 8.4, all are read in correctly without rounding or truncation.*
- *12 is an integer and gets read in as $12/10^4 = 0.0012$.*
- *123.1234 has a width of 8 and 4 decimals.*
- *All observations are displayed with 6 decimal points.*

Result:

Obs	td	
1	0.001200	**12**
2	123.000000	**123.0**
3	123.100000	**123.1**
4	123.120000	**123.12**
5	123.123000	**123.123**
6	123.123400	**123.1234**
7	0.000050	**0.00005**
8	0.000500	**0.0005**
9	0.005000	**0.005**
10	0.050000	**0.05**
11	0.500000	**0.5**

Case 2: INFORMAT for the Variable td is 8.2

SAS Codes (E322c2)

```
DATA test1;
INPUT td 8.2;
DATALINES;
12
123.0
123.1
123.12
123.123
123.1234
0.00005
0.0005
0.005
0.05
0.5
;
RUN;
PROC PRINT DATA=test1;
FORMAT td 10.6;
RUN;
```

"INPUT td 8.2;" tells SAS to read data in with a total column width of 8 and 2 decimal points.

- *Results show that all decimal points are read by SAS rather than the 2 digits specified.*
- *This phenomenon is because the 8.2 INFORMAT specification has enough columns (8 columns) to accommodate all digits of all observations. SAS is smart enough to read all the original data without alteration.*

Result:

Obs	td	
1	0.120000	**12**
2	123.000000	**123.0**
3	123.100000	**123.1**
4	123.120000	**123.12**
5	123.123000	**123.123**
6	123.123400	**123.1234**
7	0.000050	**0.00005**
8	0.000500	**0.0005**
9	0.005000	**0.005**
10	0.050000	**0.05**
11	0.500000	**0.5**

Case 3: INFORMAT for the Variable td is 7.4

SAS Codes (E322c3)

```
DATA test1;
INPUT td 7.4;
DATALINES;
12
123.0
123.1
123.12
123.123
123.1234
0.00005
0.0005
0.005
0.05
0.5
;
RUN;
PROC PRINT DATA=test1;
FORMAT td 10.6;
RUN;
```

"INPUT td 7.4;" tells SAS to read data in with a total column width of 7 and 4 decimal points.

The 7.4 INFORMAT specification offers enough columns for all observations except 123.1234, which needs 8 columns. Consequently, SAS truncates 1 column – the last decimal point – and retains all other digits to maintain maximum fidelity to the original data because of SAS's internal logic.

Result:

Obs	td	
1	0.001200	**12**
2	123.000000	**123.0**
3	123.100000	**123.1**
4	123.120000	**123.12**
5	123.123000	**123.123**
6	123.123000	**123.1234**
7	0.000050	**0.00005**
8	0.000500	**0.0005**
9	0.005000	**0.005**
10	0.050000	**0.05**
11	0.500000	**0.5**

Case 4: INFORMAT for the Variable td is 6.4

SAS Codes (E322c4)

```
DATA test1;
INPUT td 6.4;
DATALINES;
12
123.0
123.1
123.12
123.123
123.1234
0.00005
0.0005
0.005
0.05
0.5
;
RUN;
PROC PRINT DATA=test1;
FORMAT td 10.6;
RUN;
```

"INPUT td 6.4;" tells SAS to read data in with a total column width of 6 and 4 decimal points.

The 6.4 INFORMAT specification offers enough columns for all observations except 123.123 and 123.1234, which needs 7 and 8 columns, respectively.

Also, observation 0.00005 is 7 columns wide.

SAS is smart (internal logic) enough to retain all whole digits at the expense of decimal points. Consequently, 123.123 was read in as 123.12 and displayed as 123.120000 and 123.1234 was read in as 123.12 and displayed as 123.120000.

For the 0.00005 observation, the maximum 6-column is 0.0000. Consequently, 0.00005 was read in as 0.0000 and displayed as 0.000000.

Result:

Obs	td	
1	0.001200	**12**
2	123.000000	**123.0**
3	123.100000	**123.1**
4	123.120000	**123.12**
5	123.120000	**123.123**
6	123.120000	**123.1234**
7	0.000000	**0.00005**
8	0.000500	**0.0005**
9	0.005000	**0.005**
10	0.050000	**0.05**
11	0.500000	**0.5**

Case 5: INFORMAT for the Variable td is 5.4

SAS Codes (E322c5)

```
DATA test1;
INPUT td 5.4;
DATALINES;
12
123.0
123.1
123.12
123.123
123.1234
0.00005
0.0005
0.005
0.05
0.5
;
RUN;
PROC PRINT DATA=test1;
FORMAT td 10.6;
RUN;
```

"INPUT td 5.4;" tells SAS to read data in with a total column width of 5 and 4 decimal points.

The 5.4 INFORMAT specification offers enough columns for all observations except 123.12, 123.123, and 123.1234, which needs 6, 7, and 8 columns, respectively. It also affects observations 0.00005 and 0.0005.

SAS is smart (internal logic) enough to retain all whole digits at the expense of decimal points. Consequently, 123.123 was read in as 123.12 and displayed as 123.120000 and 123.1234 was read in as 123.12 and displayed as 123.120000.

0.00005 and 0.0005 were read in as 0.000 and displayed as 0.000000 and 0.000000.

Result:

Obs	td	
1	0.001200	**12**
2	123.000000	**123.0**
3	123.100000	**123.1**
4	123.100000	**123.12**
5	123.100000	**123.123**
6	123.100000	**123.1234**
7	0.000000	**0.00005**
8	0.000000	**0.0005**
9	0.005000	**0.005**
10	0.050000	**0.05**
11	0.500000	**0.5**

Case 6: INFORMAT for the Variable td is 4.**4**

SAS Codes (E322c6)

```
DATA test1;
INPUT td 4.4;
DATALINES;
12
123.0
123.1
123.12
123.123
123.1234
0.00005
0.0005
0.005
0.05
0.5
;
RUN;
PROC PRINT DATA=test1;
FORMAT td 10.6;
RUN;
```

The 4.4 INFORMAT specification offers enough columns for all observations of 12, 0.05, and 0.5.

SAS is smart (internal logic) enough to retain whole digits at the expense of decimal points for maximum fidelity to original data.

Result:

Obs	td	
1	0.001200	**12**
2	123.000000	**123.0**
3	123.000000	**123.1**
4	123.000000	**123.12**
5	123.000000	**123.123**
6	123.000000	**123.1234**
7	0.000000	**0.00005**
8	0.000000	**0.0005**
9	0.000000	**0.005**
10	0.050000	**0.05**
11	0.500000	**0.5**

Case 7: INFORMAT for the Variable td is 3.**4**

SAS Codes (E322c7)

```
DATA test1;
INPUT td 3.4;
DATALINES;
12
123.0
123.1
123.12
123.123
123.1234
0.00005
0.0005
0.005
0.05
0.5
;
RUN;
PROC PRINT DATA=test1;
FORMAT td 10.6;
RUN;
```

The 3.4 INFORMAT is improper given w is 3 and d is 4.
w is smaller than d.
w should always be greater than or equal to d.

Result:

Obs	td	
1	0.001200	**12**
2	0.012300	**123.0**
3	0.012300	**123.1**
4	0.012300	**123.12**
5	0.012300	**123.123**
6	0.012300	**123.1234**
7	0.000000	**0.00005**
8	0.000000	**0.0005**
9	0.000000	**0.005**
10	0.000000	**0.05**
11	0.500000	**0.5**

Special Cases – Numerical Data Containing $, %, ( ), and/or Comma (,)

You can have SAS to read 12,345,678 with the **INFORMAT** statement of COMMA10. The reason we use **Comma10.** is due to the fact that 12,345,678 has a total of 10 columns, and the number is separated with the comma. You can use **11.** or any other number greater than 10 to accomplish the reading, too. The **COMMAw.d** enables the reading of such special case data correctly.

SAS Codes (E322sc)

```
DATA new;
INPUT data_comma comma10. data_dollar comma12. data_percent
comma5.  data_parenthesis comma7.;
DATALINES;
12,345,678 $12,345,678 12.6% (12.5)
;
PROC PRINT;
RUN;
;
```

INFORMAT comma w.
comma10. reads data with a maximum of 10 columns width and containing , $, %, ( ) symbols. The results are that these symbols are removed.

Result:

Obs	data_comma	data_dollar	data_percent	data_parenthesis
1	12345678	12345678	12.6	-12.5

3.2.3 Date/Time/Datetime

Be sure to pay close attention to date/time variables as they can be expressed in a wide range of formats. You may encounter a lot of frustration trying to fully address all techniques and methods dealing with date, time, or datetime data.

The date and time variables can help you to identify relationships and trends in your data. When you see date and time variables, in addition to considering trending, you should also consider the following possibilities:

- Day of week patterns and effects
- Weekday and weekend effects and variation
- Time of day – daily patterns
- Number of days between two dates effect – effects from days and time-lapsed.

Below are examples illustrating usages of different date and time **INFORMAT** statements.

Example 1: MMDDYY8.

Reading dates written as **mm/dd/yy** or **mm.dd.yy** or **mm-dd-yy**

SAS Codes (E323e1)

```
DATA test18;
FORMAT mydate mmddyy8.;
INPUT mydate mmddyy8.;
DATALINES;
12/23/99
01.09.89
02-28-19
;
PROC PRINT data=test18;
RUN;
```

Result:

The SAS System

Obs	mydate
1	12/23/99
2	01/09/89
3	02/28/19

Example 2: MMDDYY10.

Reading dates written as **mm/dd/yyyy** or **mm.dd.yyyy** or **mm-dd-yyyy**

SAS Codes (E323e2)

```
DATA test18;
FORMAT mydate mmddyy10.;
INPUT mydate mmddyy10.;
DATALINES;
12-08-2015
12/08/2015
05/25/2015
;

PROC PRINT data=test18;
RUN;
```

Result:

The SAS System

Obs	mydate
1	12/08/2015
2	12/08/2015
3	05/25/2015

Example 3: DDMMMYY7.

Reading dates in the form of **ddMMMyy**

SAS Codes (E323e3)

```
DATA test18;
FORMAT mydate ddmmmyy7.;
INPUT mydate ddmmmyy7.;
DATALINES;
02sep18
21JAN19
16apr78
;
PROC PRINT data=test18;
RUN;
```

Result:

The SAS System

Obs	mydate
1	12/08/2015
2	12/08/2015
3	05/25/2015

Example 4: DATE9.

Reading dates in the form of **date9.**

SAS Codes (E323e4)

```
DATA test18;
FORMAT mydate date9.;
INPUT mydate date9.;
```

```
DATALINES;
12sep1978
05jan2019
16apr1978
;
PROC PRINT data=test18;
RUN;
```

Result:

The SAS System

Obs	mydate
1	12SEP1978
2	05JAN2019
3	16APR1978

Example 5: TIMEw.
Reading time in the form of **hhMMss**

SAS Codes (E323e5)

```
DATA test18;
FORMAT mytime time8.;
INPUT MyTime time8.;
DATALINES;
09:30:26
14:36:28
08:02:03
;
RUN;
PROC PRINT data=test18;
RUN;
```

Result:

The SAS System

Obs	mytime
1	9:30:26
2	14:36:28
3	8:02:03

Example 6: ANYDTDTMw.

Reading date and time in the form of **ANYDTDTM20**.

SAS Codes (E323e6)

```
DATA test6;
FORMAT mydatetime mdyampm.;
INPUT MyDateTime ANYDTDTM20.;
DATALINES;
12/09/19 9:23:18
8/12/19 10:23:18
01/08/17 23:07:12
;
PROC PRINT data=test6;
RUN;
```

Result:

The SAS System

Obs	mydatetime
1	12/9/2019 9:23 AM
2	8/12/2019 10:23 AM
3	1/8/2017 11:07 PM

- **INFORMAT** statements tell SAS how to read data and impact data precision.
- **FORMAT** statements tell SAS how data should be displayed in the output and do not affect data precision.

You can specify variable types as part of an INPUT statement, as illustrated in all the examples so far in this chapter (Approach 1). You can also specify variable types through a separate statement beginning with the **INFORMAT** statement (Approach 2).

Approach 1

```
Data New;
INPUT myvar1 $3.  myvar2 $6. myvar3 4. myvar4  8.  mydate
ddmmyyyy10. ;
```

Approach 2

```
Data New;
INFORMAT
myvar1   $3.
myvar2   $6.
```

```
myvar3      4.
myvar4       8.
mydate      ddmmyyyy10.
;
INPUT     myvar1   myvar2   myvar3   myvar4   mydate;
```

Specify Similar Variable Types at Once

When you have a set of similar variables, you can specify their types in many ways. For example, your three variables **CensusRname, CensusDname,** and **StateAbb** are all character variables. They have a maximum width of two. You can specify them in two different methods, as illustrated below.

1st Method: Specifying all 3 variable names separately

```
INPUT CensusRname $2.  CensusDname $2.   StateAbb $2.    CensusRid
CensusDid  StateFIPS  MaleP FemalP;
```

2nd Method: Specifying all 3 together at once

You must use the parentheses to group similar variables together first. Once grouped, you then specify the type, also in parentheses.

```
INPUT (CensusRname  CensusDname  StateAbb) ($2.)  CensusRid
CensusDid  StateFIPS  MaleP FemalP;
```

3.2.4 Missing Value

SAS handles missing value differently for character data, numerical data, and date/time data. Missing numerical data are indicated by a period (.) symbol. Missing a character value is represented by a blank in the output data.

See the examples below.

Example

SAS Codes (E324a)

```
DATA new;
INPUT row_number name $  age;
DATALINES;
1 Marsha    35
2 Toby      27
3 Leo       .
4 Josh      22
5 Beth      .
6 .         26
7 .         57
```

```
8 Don        29
;
PROC PRINT DATA=new;
RUN;
```

Result:

Obs	row_number	name	age
1	1	Marsha	35
2	2	Toby	27
3	3	Leo	.
4	4	Josh	22
5	5	Beth	.
6	6		26
7	7		57
8	8	Don	29

3.2.5 SAS Date

Our current Gregorian calendar uses the year of Jesus' birth as year zero and counts forward from this reference year. For example, year 2022 AD means 2022 years after this reference year. SAS has its own date counting system. SAS's reference date is January 1, 1960. This date is referred to as day zero. Any date after 1/1/1960 will be a positive number. Dates prior to 1/1/1969 will be negative. See the illustrations below.

Example

SAS Codes E325

```
DATA testB;
FORMAT mydate MMDDYY8.;
```

*The above format statement requests SAS to display the mydate varible in testB data in mmddyy8. format (shown on the left side of the Result);

```
INPUT mydate MMDDYY8. ;
DATALINES;
12/31/59
1/1/60
1/2/60
12/28/19
8/1/49
;
```

```
PROC PRINT data=testB;
RUN;

DATA testA;
INPUT MyDateA  MMDDYY8. ;
DATALINES;
12/31/59
1/1/60
1/2/60
12/28/19
8/1/49
;
```

***No format is given to the MyDateA variable in TestA data, it defaults to the SAS date format (shown on the right side of the Result);**

```
PROC PRINT data=testA;
RUN;
```

Result:

The SAS System

Obs	mydate
1	12/31/59
2	01/01/60
3	01/02/60
4	12/28/19
5	08/01/49

SAS Date

Obs	mydate
1	-1
2	0
3	1
4	21911
5	-3805

3.2.6 Summary

Understanding variable types is very critical to input your data into SAS correctly. While a variable name can be as long as 32 characters, the default observation length varies depending on the variable type. For a character variable, the default length for your observation is only eight characters wide (8 columns). Any character observation that has a length greater than eight will be truncated unless specified otherwise with a customized INFORMAT.

There is no default length for numerical variable observations in the **INFORMAT** form **w.d**.

The default lengths for date, time, and datetime vary. Pay close attention when dealing with such information.

3.3 EMBEDDING YOUR DATA WITH YOUR CODE FILE

When your data is limited (e.g., less than 100 rows), you can conveniently embed your data into your SAS code file by listing it following the "DATALINES" statement. Examples in Section 3.2 have demonstrated such embedding practices.

Space (number of blanks) used to separate observations for different variables does not need to be uniform throughout your data. SAS counts such separation blanks as 1 column regardless of the number of columns you may actually choose to use.

Example 3.3a: Initial test

SAS Codes (E33a)

```
DATA exer_embed_1;
INPUT RegionName $
RegionID $    Pop2010
Pop2011     Pop2012;
DATALINES;
Northeast    N20    55317240
        55318430     55380645
Midwest      M30
        66927001     66929743
        66974749
South S40    14555744
        14563045     14867066
West W50    71945553
        71946887     72103625
;
PROC PRINT
DATA=EXER_EMBED_1;
RUN;
```

DATA creates a new dataset called EXER_EMBED_1 and stores it in the default WORK library.

INPUT lists variables RegionName and RegionID as character variables, and Pop2010, Pop2011, and Pop2012 as numerical type.

DATALINES tells SAS actual data is embedded and is next.

4 rows of observations are listed. Blank spaces between observations among variables vary and are not uniform. However, for a given observation (row), each variable is separated by no less than 1 blank space.

PROC PRINT DATA prints the just created data EXER_EMBED_1 on the screen.

RUN kicks off the PROC PRINT procedure.

The printed result shows the actual data created by the DATA step.

1st row of RegionName is missing the last (9th) character.

This is because the INFORMAT for the RegionName did not specify the length of potential variables. So, default length, which is the first 8 columns, is read. Anything longer than 8 are truncated.

Result:

Obs	RegionName	RegionID	Pop2010	Pop2011	Pop2012
1	Northeas	N20	55317240	55318430	55380645
2	Midwest	M30	66927001	66929743	66974749
3	South	S40	14555744	14563045	14867066
4	West	W50	71945553	71946887	72103625

Example 3.3b: Overcoming the incomplete reading of row 1 "Northeast" as "Northeas"

The observation of "Northeast" under the variable RegionName has nine characters. Without a customized INFORMAT, by default, SAS only reads the first eight characters. We can use the following INFORMAT to overcome this issue.

SAS Codes (E33b)

```
DATA exer_embed_1;
INPUT RegionName $9.
RegionID $    Pop2010
Pop2011    Pop2012;
DATALINES;
Northeast    N20    55317240
       55318430    55380645
Midwest      M30    66927001
       66929743    66974749
South S40    14555744
       14563045    14867066
West W50     71945553
       71946887    72103625
;
PROC PRINT
DATA=EXER_EMBED_1;
RUN;
```

To overcome the issue, we specify the length of the RegionName variable with $9.

From the result, you can tell that "Northeast" was read in correctly. However, other rows are messed up.

When the INFORMAT for RegionName $9. is used, SAS reads 9 columns regardless of whether these 9 columns are separated by blanks or not. And it treats all these 9 columns as observation under the variable RegionName. In addition, the number of blank spaces used in the embedded data to separate observations for different variables is always counted as 1 blank space (e.g., blank spaces between Midwest and M30 are more than 1, but counted as 1).

Once the first 9 columns are read, SAS will read from the 10th column until it encounters a blank space before stopping. The reason a blank space will stop SAS in this example is that the INFORMAT for RegionID $ does not have a length (columns) specified. Blank space is the default variable separator.

Result:

Obs	RegionName	RegionID	Pop2010	Pop2011	Pop2012
1	Northeast	N20	55317240	55318430	55380645
2	Midwest M	30	66927001	66929743	66974749
3	South S40	14555744	14563045	14867066	.

Example 3.3c: Overcoming the misreading of the 3rd- and 4th-row observations

SAS Codes (E33c)

/*I use the informat :$9. to read up to 9 columns for the variable RegionName or until a blank space is encountered"/

```
DATA exer_embed_1;
INPUT RegionName :$9.  RegionID $
Pop2010     Pop2011     Pop2012;
DATALINES;
Northeast    N20    55317240
      55318430    55380645
Midwest      M30    66927001
      66929743    66974749
South S40    14555744    14563045
      14867066
West W50     71945553    71946887
      72103625
;
PROC PRINT DATA=EXER_EMBED_1;
RUN;
```

To correct all the issues, the INFORMAT specification is changed to RegionName :$9. The addition of the colon : symbol tells SAS to read up to 9 columns if SAS has not encountered a blank space prior to reach the 9th column. So, if SAS encounters a blank space before reaching the 9th column, it will stop. Otherwise, it will read up to the 9th column and treat what are read under the variable of RegionName.

Based on the result shown, this approach works well here. All data are read correctly.

Result:

Obs	RegionName	RegionID	Pop2010	Pop2011	Pop2012
1	Northeast	N20	55317240	55318430	55380645
2	Midwest	M30	66927001	66929743	66974749
3	South	S40	14555744	14563045	14867066
4	West	W50	71945553	71946887	72103625

3.4 INPUTTING DATA FROM AN EXTERNAL DATA FILE TO SAS

Reading data from an external data file is the most common scenario you will encounter in SAS. Data can be read from:

- SAS data (.sas7bdat)
- Other statistical package datasets such as STATA's .dta and SPSS's .sav
- Spreadsheet files like the MS Excel (.xls, xlsx)
- Database files (.DBF)
- Text files (.txt) such as comma-separated files (.csv), space- or other character- or symbol-separated text files (.txt), fixed column width text files (.txt), and others (.txt).

To read an external data file, you need to tell SAS where to find the dataset and how variables are arranged.

3.4.1 Inputting Data Stored in SAS Data Format (.sas7bdat)

SAS Codes (E341) to read the myscore.sas7bdat file

```
LIBNAME my 'c:\sas_exercise';
DATA new;
SET my.myscore;
RUN;
PROC PRINT DATA=new;
RUN;
```

LIBNAME defines a library named my representing "c:\sas_exercise."

DATA creates a new dataset called new.sas7bdat in the WORK library.

SET instructs SAS to read the myscore.sas7bdat from the my library.

RUN kicks off the DATA step.

PROC PRINT DATA prints the data named new.sas7bdat in the WORK library.

RUN kicks off the PROC PRINT procedure.

3.4.2 Reading a Comma (,)-Separated Text File (.csv)

SAS Codes (E342) to read the myscore.csv file

```
DATA myTestData;
INFILE 'C:\sas_exercise\
myscore.csv' DLM=','
FIRSTOBS=2 OBS=76 DSD
MISSOVER;
INPUT    StateName $    StateID
RoadID $       Year      AADT;
RUN;
PROC PRINT DATA=myTestData;
RUN;
```

DATA creates a new dataset called myTestData in the WORK library.

The original data used to create the new dataset is specified by the INFILE command. The file location is "c:\sas_exercise" directory, and the actual file is myscore.csv. DLM stands for data delimiter. DLM is the , comma here because the original file is a csv (comma-separated value) file.

FIRSTOBS = 2 tells SAS to start to read data from the 2nd row (2nd observation). Often the 1st row is the title row.

OBS = 76 tells SAS to stop reading data after the 76th row is read.

DSD instructs SAS to handle 3 things:

1: Treat two consecutive delimiters (, ,) as a missing value.

2: Remove quotation marks from character values and treat everything within a quotation mark together as a single observation for the variable (e.g., "AB, C", "XYZ" will be treated as 2 observations not 3. The "AB, C" is treated as 1 and XYZ is the other one).

3: Assume data separator is the , comma.

Additional options as listed below are often used with the INFILE statement to overcome certain data issues as explained below.

FLOWOVER is the default way of SAS reading data. If there is a missing value, SAS simply reads the next value and assigns it to the variable listed in the INPUT statement.

STOPOVER instructs SAS to stop reading data when it reaches the end of a data line even though more variables listed in the INPUT statement are not provided with values.

MISSOVER prevents SAS from going to the next data line (row) if it can't locate values with the current dateline (row) for all variables listed in the INPUT statement. SAS assigns a missing value for variables without values (missing observation).

TRUNCOVER instructs SAS to assign a value to a variable even if the value is shorter (width) than what is specified in the INPUT INFORMAT. In the event that SAS reaches the end of a data line (row) and there are still variables without assigned values, missing values are assigned to these variables.

The INPUT command lists all variables and associated INFORMAT tied with the myscore.csv file. StateName is a character variable. StateID is a numeric variable. RoadID is a character variable. Year and AADT are all numerical variables.

RUN kicks off the DATA step.

3.4.3 Reading a Tab Key-Separated Text File (.txt or .tab)

SAS Codes (E343) to read the myscore.txt file

```
DATA myTestData;
INFILE    'C:\sas_exercise\
myscore.csv'   DLM='09'x
FIRSTOBS=2   MISSOVER;
INPUT    StateName $        StateID
RoadID $       Year        AADT;
RUN;
PROC PRINT DATA=mytestdata;
RUN;
```

Observations in the raw external text file are separated by the Tab key (vs. comma). So, the option for "DLM = '09'x" is used. All others are similar to E342.

3.4.4 Reading a Vertical Bar Symbol (|)-Separated Text File (.txt)

SAS Codes (E344) to read the myscore.txt file

```
DATA myTestData;
INFILE     'C:\sas_exercise\myscore.
txt'   DLM='|'   FIRSTOBS =2   OBS=50
MISSOVER ;
INPUT StateName    $        StateID
RoadID $        Year        AADT;
RUN;
PROC PRINT DATA=mydata;
RUN;
```

Observations in the raw external text file are separated by the vertical bar (|) key (vs. comma). So the option for "DLM = '|'" is used. All others are similar to E342.

3.4.5 Reading a Text Data File with Fixed Starting and Ending Column Locations and Widths

Example

Column 1–10	StateName
Column 11–12	StateID
Column 13–23	RoadID
Column 24–27	Year (YYYY)
Column 28–34	AADT (Annual Average Daily Traffic)

SAS Codes (E345) to read the volum8.txt file

```
DATA mydata;
INFILE      'c:\sas_exercise\
school_project\volume8.txt'
TRUNCOVER;
INPUT
StateName $ 1-10
StateID 11-12
RoadID $ 13-23
Year 24-27
AADT 28-34
;
RUN;
PROC PRINT DATA=mydata;
RUN;
```

ATTENTION

The command **SET** is only used to read an existing SAS data file that has the extension of .sas7bdat. The command **INPUT** reads raw data from data files other than the .sas7bdat type.

The DATA step reads an external txt file with a rigid column arrangement for all variables.

DATA command creates a new SAS data file called mydata. The file is based on the volume8.txt file, which is specified by the INFILE command with a full directory.

The TRUNCOVER option instructs SAS to assign the value to the variable even if the value is shorter (width) than what is specified in the INPUT INFORMAT. In the event that SAS reaches the end of a data line and there are still variables without assigned values, missing values are assigned to these variables.

INPUT lists all variables with their rigid INFORMAT.

StateName is a character variable occupying columns 1 to 10.

StateID is a numerical variable occupying columns 11 and 12.

RoadID is another character variable occupying columns from 13 to 23.

Year is a numerical variable occupying columns 24 to 27.

AADT is a numerical variable occupying columns 28 to 34.

"RUN;" kicks off the DATA step.

PROC PRINT prints the newly created dataset through DATA = mydata on the screen.

"RUN;" kicks off the PROC PRINT procedure.

3.5 IMPORTING DATA TO SAS

You can import data into SAS. The difference between reading data into SAS and importing is that when SAS imports data, it automatically checks variable types (character, numerical, date) by examining the first 20

Importing Data

The most significant advantage of importing data is its auto-check of variable type.

rows of the dataset. This means that SAS will automatically detect variable type when importing option is used.

3.5.1 Importing an Excel Data File

SAS Codes (E351) to import an Excel data file

```
PROC IMPORT OUT=
WORK.mytest1
           DATAFILE=
"c:\exercise\test1.
xls"
  DBMS=EXCEL REPLACE;
 RANGE="Sheet1$";
 GETNAMES=YES;
 MIXED=NO;
 SCANTEXT=YES;
 USEDATE=YES;
 SCANTIME=YES;
RUN;
```

PROC IMPORT procedure invokes SAS's data importing function.

The OUT = WORK.mytest1 tells SAS to generate a new dataset called mytest1.sas7bdat (default SAS extension can be omitted) and put it in the WORK library.

DATAFILE specifies that the text1.xls file in the folder of c:\exercise is to be imported.

The DBMS (database management systems) command tells SAS that the test1.xls data is an Excel file.

The REPLACE instructs SAS to replace any existing files in the WORK library having the name of mytest1.sas7bdat with the new one.

RANGE tells SAS to import Sheet1 of test1.xls through the "Sheet1$" specification. Pay attention to the $ added to the sheet name.

"GETNAMES = YES" tells SAS to read the title of the columns in the Excel file.

MIXED controls variable INFORMATs. By default, SAS examines the first 8 rows of a column to determine variable INFORMAT. If your data are consistent, this works well. The Option "YES" tells SAS that if multiple formats are found for a variable, then the variable should be treated as a character variable.

SCANTEXT is applicable to text variables (columns). SCANTEXT = YES scans the column for the longest string and uses its length as the column width.

USEDATE = YES instructs SAS to use the date format setting in the original Excel file. (Additional Information - USEDATE = NO instructs SAS to read your date as a string.)

SCANTIME = YES instructs SAS to scan variables for time-specific formats.

RUN kicks off the PROC IMPORT procedure.

3.5.2 Importing a Comma (,)-Separated Data File

SAS Codes (E352) to import a Comma (,)-separated text data file

```
PROC IMPORT OUT= WORK.scoreData
            DATAFILE= "c:\exercise\test2.csv"
             DBMS=CSV REPLACE;
     GETNAMES=YES;
     DATAROW=2;
RUN;
```

PROC IMPORT procedure invokes SAS's data importing function.
The OUT = WORK.scoreData tells SAS to generate a new dataset called scoreData.sas7bdat (default SAS extension can be omitted) and put it in the WORK library.
DATAFILE specifies that the text2.csv file in the folder of c:\exercise is to be imported.
The DBMS (database management systems) command tells SAS that the test2.csv is a comma-separated value format. The REPLACE instructs SAS to replace any existing files in the WORK library, having the name of scoredata.sas7bdat with the new one.
"GETNAMES = YES" tells SAS to read the title of the columns (variables).
DATAROW = 2 tells SAS that actual data starts at row2.
RUN kicks off the PROC IMPORT procedure.

3.5.3 Importing a Tab-Delimited File

SAS Codes (E353) to import a Tab-delimited file

```
PROC IMPORT OUT= WORK.TestData7
            DATAFILE= "C:\exercise\test_7.txt"
            DBMS=TAB REPLACE;
     GETNAMES=YES;
     DATAROW=2;
RUN;
```

PROC IMPORT procedure invokes SAS's data importing function.
The OUT = WORK.TestData7 tells SAS to generate a new dataset called TestData7.sas7bdat (default SAS extension can be omitted) and put it in the WORK library.
DATAFILE specifies the test_7.txt file in the folder of c:\exercise is to be imported.
The DBMS (database management systems) command tells SAS that the test_7.txt is a TAB-separated value format.
The REPLACE instructs SAS to replace any existing files in the WORK library having the name of TestData7.sas7bdat with the new one.
"GETNAMES = YES" tells SAS to read the title of the columns (variables).
DATAROW = 2 tells SAS that actual data starts at row2.
RUN kicks off the PROC IMPORT procedure.

3.5.4 Importing a Period (.)-Delimited File

SAS Codes (E354) to import a Period (.)-delimited file

```
PROC IMPORT OUT= myPfile
            DATAFILE= "C:\exercise\final52.csv"
            DBMS=DLM REPLACE;
     DELIMITER='00'x;
     GETNAMES=YES;
     DATAROW=2;
RUN;
```

PROC IMPORT procedure invokes SAS's data importing function.

The OUT = myPfile tells SAS to generate a new dataset called myPfile.sas7bdat (default SAS extension can be omitted) and put it in the WORK library.

DATAFILE specifies that the final52.csv file in the folder of c:\exercise is to be imported.

The DBMS (database management systems) command tells SAS that the final52.csv is a delimited file. The REPLACE instructs SAS to replace the to-be-created myPfile.sas7bdat file in the WORK library if there is already one there.

DELIMITER = '00'x tells SAS that the actual delimiter used is the . period.

DATAROW = 2 tells SAS that actual data starts at row2.

"GETNAMES = YES" tells SAS to read the title of the columns (variables).

RUN kicks off the PROC IMPORT procedure.

3.5.5 Importing a DBF File

SAS Codes (E355) to import a DBF file

```
PROC IMPORT OUT= WORK.exer12
DATAFILE= "c:\exercise\test.dbf"
             DBMS=DBF REPLACE;
     GETDELETED=NO;
RUN;
```

PROC IMPORT procedure invokes SAS's data importing function.

The OUT = WORK.exer12 tells SAS to generate a new dataset called exer12.sas7bdat (default SAS extension can be omitted) and put it in the WORK library.

DATAFILE specifies that the test.dbf file in the folder of c:\exercise is to be imported.

The DBMS (database management systems) command tells SAS that the test.dbf is a DBF file.

The REPLACE instructs SAS to replace any existing files in the WORK library having the name of exer12.sas7bdat with the new one.

"GETDELETED = NO" instructs SAS not to read and write rows to the SAS dataset (exer12) that are marked for deletion but have not been purged in the original DBF file.

RUN kicks off the PROC IMPORT procedure.

3.5.6 Importing an SPSS Data File

SAS Codes (E356) to import an SPSS data file

```
PROC IMPORT OUT= mydata1
            DATAFILE= "C:\exercise\fh2.sav"
            DBMS=SPSS REPLACE;
RUN;
```

PROC IMPORT procedure invokes SAS's data importing function.
The OUT = mydata1 tells SAS to generate a new dataset called mydata1.sas7bdat (default SAS extension can be omitted) and put it in the WORK library.
DATAFILE specifies fh2.sav file in the folder of c:\exercise is to be imported.
The DBMS (database management systems) command tells SAS that the fh2.sav is a SPSS file. The REPLACE instructs SAS to replace any existing files in the WORK library having the name of mydata1.sas7bdat with the new one.
RUN kicks off the PROC IMPORT procedure.

3.5.7 Importing a STATA Data File

SAS Codes (E357) to import a STATA data file

```
PROC IMPORT OUT= WORK.MyStata8
            DATAFILE= "C:\exercise\test_8.dta"
            DBMS=STATA REPLACE;
RUN;
```

PROC IMPORT procedure invokes SAS's data importing function.
The OUT = WORK.MyStata8 tells SAS to generate a new dataset called MyStata8.sas7bdat (default SAS extension can be omitted) and put it in the WORK library.
DATAFILE specifies that test_8.dta file in the folder of c:\exercise is to be imported.
The DBMS (database management systems) command tells SAS that the test_8.dta is a STATA file. The REPLACE instructs SAS to replace any existing files in the WORK library having the name of MyStata8.sas7bdat with the new one.
RUN kicks off the PROC IMPORT procedure.

3.6 READING OTHER LESS STRUCTURED DATA

When data are not well organized in rows and columns, additional steps may be needed to feed such data into SAS.

Example 3.6

Data:
A flat text file below contains data for different airport gates and the numbers of male and female passengers the gates served in a year.

```
IAD9801   2001   125678   136791   IAD011   2002   145678 196791 IAD081
2002   .  IAD6721 2004   456981   679812IAD6721   2005 .   789651
```

Task: Read the data into SAS

The above data organization does not follow the standard one observation per line database structure. Each line has multiple observations. In cases like this, you can use the following statement to read such data by deploying the double trailing @@ symbol.

SAS Codes (E36)

```
DATA exercise_1;
INPUT  GateID $     Year   MaleP
FemaleP   @@    ;
DATALINES;
IADO801    2001      125678
136791   IAD0117  2002  145678
196791
IAD0812 2002   .   .
IAD6721 2004  456981     679812
IAD6721     2005    .  789651
;
PROC PRINT DATA=exercise_1;
RUN;
```

DATA creates a new SAS dataset exercise_1 in the WORK library. The INPUT lists all variables with their INFORMATs, where GateID is a character variable ($), and Year, MaleP, and FemaleP are all numerical variables.

@@ instructs SAS to continue to read data once it reaches the end of a data line.

DATALINES signals the starting of actual data.

Obviously, the observations here do not follow per row standard specification. Instead, it continues. This is why we use the @@ in the INPUT statement.

The PROC PRINT DATA = exercise_1 instructs SAS to print the exercise_1.sas7bdat on the screen.

RUN kicks off the PROC PRINT procedure.

The result appears correct.

The consecutive trailing @@ at the end of the **INPUT** statement instructs SAS to continue to read the data lines by following the sequence of **INPUT** variables.

Below is the result of reading the above data.

Result:

Obs	GateID	Year	MaleP	FemaleP
1	IADO801	2001	125678	136791
2	IAD0117	2002	145678	196791
3	IAD0812	2002	.	.
4	IAD6721	2004	456981	679812
5	IAD6721	2005	.	789651

3.7 LISTING INPUT VARIABLES ALTERNATIVELY

You can use a single hyphen (-) to specify a range of variables that share a common prefix and a sequential set of numerical suffixes. For example, **varl, var2, var3, var4, var4, var 6** can be specified as **varl - var6.**

SAS Code (E37a)

```
DATA new;
INPUT  varl - var6  ;
```

is the same as

```
DATA new;
INPUT  varl  var2  var3  var4
var5  var6  ;
```

DATA creates a new SAS dataset called new in the WORK library.

INPUT var1 - var6 tells SAS that there are 6 variables having the same prefix var and different suffix from 1 to 6: var1, var2, var3, var4, var5, and var6.

The shortened specification is the same as the regular specification with the INPUT command of INPUT var1 var2 var3 var4 var5 var6.

You can also use the _NUMERIC_, _CHARACTER_, and _ALL_ keywords to specify variables of certain types (numeric or character) or all types, as illustrated below.

SAS Code (E37b)

```
PROC PRINT DATA=test;
VAR  _CHARACTER_;
RUN;
```

PROC PRINT DATA = test instructs SAS to output the test data in the WORK library on a screen.

The VAR_CHARACTER_further specifies that SAS should only output character variables.

RUN kicks off the PROC PRINT procedure.

4

Data Manipulation

In this chapter, data manipulation refers to a broad range of data activities, including exporting SAS data out to other formats; subsetting, dividing, or splitting a dataset into smaller datasets; merging datasets vertically; joining dataset left/right; and creating new data variables through a wide range of methods.

Be sure to pay close attention to all the examples to understand the command, statement, logic, and different approaches available to accomplish a given task.

4.1 EXPORTING A DATA FILE

You can export data from one format to another. For example, you can export a SAS dataset (.sas7bdat) to a comma-separated value file (e.g., csv), an Excel (.xlsx or .xls), or other delimited (e.g., .txt) files.

4.1.1 Exporting a SAS Data File (.sas7bdat) to a Comma-Delimited File

Method 1

SAS Codes (E41a1)

```
LIBNAME mydata 'c:\sas_exercise';
DATA region2;
SET mydata.regionbb;
RUN;
PROC EXPORT DATA=region2
OUTFILE = 'c:\sas_exercise\NWregion1.csv'
DBMS=csv
REPLACE;
RUN;
```

LIBNAME defines a library called mydata representing the folder (directory) of c:\sas_exercise.

DATA creates a new SAS dataset called region2 in the default WORK library.

SET tells SAS that the new dataset region2 is based on an existing SAS data called regionbb.sas7bdat stored in the mydata library (folder of c:\sas_exercise).

The mydata.regionbb uses the two-layer SAS data name convention, where mydata is the library and regionbb is the SAS data file name.

RUN kicks off the DATA step.

PROC EXPORT invokes the export function. DATA=region2 specifies the data to be exported is the newly created region2 data in the WORK library.

OUTFILE specifies the directory path (c:\sas_exercise) and the new file to be created named NWregion1.csv.

The DBMS (database management systems) command tells SAS that the NWregion1.csv should be a comma-separated value file (csv) file. REPLACE instructs SAS to replace any existing file having the name of NWregion1.csv in the folder of c:\sas_exercise with the newly created NWregion1.csv file.

RUN kicks off the PROC EXPORT procedure.

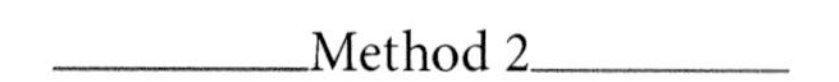

Method 2

SAS Codes (E41a2)

```
LIBNAME mydata 'c:\sas_exercise';
PROC EXPORT DATA=mydata.regionbb
OUTFILE = 'c:\sas_exercise\
NEWregion2.csv'
DBMS=csv
REPLACE;
RUN;
```

LIBNAME defines a library called mydata representing the folder (directory) of c:\sas_exercise.

PROC EXPORT invokes the export function. DATA=mydata.regionbb specifies the data to be exported is called regionbb.sas7bdat stored in the library of mydata. Here, mydata.regionbb uses the two-layer SAS data file naming convention.

OUTFILE specifies the directory path (c:\sas_exercise) and the new file to be created named NEWregion2.csv.

The DBMS (database management systems) command tells SAS that the NEWregion2.csv should be a comma-separated value (csv) file.

REPLACE instructs SAS to replace any existing file having the name of NEWregion2.csv in the folder of c:\sas_exercise with the newly created NEWregion2.csv file.

RUN kicks off the PROC EXPORT procedure.

4.1.2 Exporting a SAS Data File (.sas7bdat) to an Excel File

SAS Codes (E41b)

```
LIBNAME mydata 'c:\sas_exercise';
DATA mydata.regionK;
SET mydata.regionbb ;
RUN;
```

LIBNAME defines a library called mydata representing the folder (directory) of c:\sas_exercise.

DATA mydata.regionK creates a new SAS dataset called regionK to be stored in the mydata library.

```
PROC EXPORT DATA=mydata.
regionK
OUTFILE = 'c:\sas_exercise\
NEWregion2.xls'
DBMS=EXCEL
REPLACE;
RUN;
```

SET mydata.regionbb tells SAS that the new dataset regionK is based on an existing SAS data called regionbb.sas7bdat stored in the mydata library (folder of c:\sas_exercise).

The mydata.regionbb uses the two-layer SAS data naming convention where mydata is the library and regionbb is the SAS data file name.

RUN kicks off the DATA step.

PROC EXPORT invokes the export function. DATA=mydata.regionK tells SAS that the to-be-exported dataset is called region.sas7bdat and stored in the mydata library.

OUTFILE specifies the directory path (c:\sas_exercise) and the new file to be created named NEWregion2.xls.

The DBMS (database management systems) tells SAS that the NEWregion2.xls should be an Excel file.

REPLACE instructs SAS to replace any existing file having the name of NEWregion2.xls in the folder of c:\sas_exercise with the newly created NEWregion2.xls file.

RUN kicks off the PROC EXPORT procedure.

4.2 SPLITTING/SUBSETTING A DATA FILE

For a lot of reasons, you may want to split your data file into multiple files. You can split a file by the number of rows (observations) or by a particular data attribute.

For example, your data has 2.6 million rows of observations. You can split the data into three separate files with the first file having 900,000 rows, the second one possessing 950,000 rows, and the last one containing 750,000 rows.

You can also split a file based on a particular data attribute. For example, if your data contain a month variable, you can split your data into 12 monthly files by using the month as the attribute.

Once your data is split, you can use the export function to create new files in whatever format you desire (e.g., .csv, .xls, .xlsx, .dbf).

4.2.1 Dividing a SAS Data File by Number of Rows.

If you plan to split a file based on the number of rows it has, you should first determine the total number of rows your file has. The SAS procedure used for this is the **PROC CONTENTS** statement.

Example: 4.2.1

Data: split2.csv

Task: Create three separate files from the original file where the first file contains 500 rows, the second one 300 rows, and the last one 200 rows.

Steps:

a: Determine the number of rows the split2.csv has

You can use the **PROC CONTENTS** statement to see the number of rows (observations), as illustrated below.

SAS Codes (E421a)

```
DATA splitData;
INFILE 'c:\sas_exercise\split2.csv' DLM=',' FIRSTOBS=2   ;
INPUT ID region $     StateID $ population ;
RUN;
PROC CONTENTS DATA = work.splitData;
RUN;
```

The SAS System

The CONTENTS Procedure

Data Set Name	WORK.SPLITDATA	Observations	1000
Member Type	DATA	Variables	4
Engine	V9	Indexes	0
Created	Friday, December 13, 2019 01:34:05 PM	Observation Length	32
Last Modified	Friday, December 13, 2019 01:34:05 PM	Deleted Observations	0
Protection		Compressed	NO
Data Set Type		Sorted	NO
Label			
Data Representation	WINDOWS_32		
Encoding	wlatin1 Western (Windows)		

DATA creates a new SAS dataset called splitData.sas7bdat in the default WORK library (only one-layer name is used. No library is specified so it defaults to the WORK library).

INFILE tells SAS that the to-be-created new splitData is based on the external split2.csv data file stored in the c:\sas_exercise folder.

DLM=',' tells SAS that data delimiter in the split2.csv is comma.

FIRSTOBS=2 tells SAS to start to read the split2.csv data from row 2 (possibly, the 1st row is the header row in the split2.csv file).

INPUT lists variables with their INFORMATs. Both the region and StateID are character variables evidenced by the $ INFORMAT, and population is a standard numerical variable.

RUN kicks off the DATA step.

PROC CONTENTS DATA=WORK.splitdata invokes SAS PROC procedure of CONTENTS where a host of parameters describing contents of the splitdata will be generated. Here, a two-layer naming convention is adopted. WORK is the library, and splitdata is the datafile name.

RUN kicks off the PROC CONTENTS procedure.

The result shows that there are 1000 (observations) rows of data counted from row 2 as instructed by the DATA step INFILE FIRSTOBS=2 command.

b: Review the result as illustrated above: The number of "**observations**" (rows) is 1000.

SAS Codes (E421b)

```
DATA sd1;
INFILE 'c:\sas_exercise\split2.csv' DLM=',' FIRSTOBS=1 obs=500;
INPUT ID region $     StateID $ population ;
RUN;
PROC EXPORT DATA=sd1 OUTFILE ='c:\sas_exercise\sd1.csv'
DBMS =dlm REPLACE;
delimiter=',';
RUN;
```

DATA sd1 creates a new SAS dataset named sd1.sas7bdat and stored in the default WORK library.

INFILE tells SAS that the to- be-created new sd1.sas7bdat is based on the external split2.csv data file stored in the c:\sas_exercise folder.

DLM= ',' tells SAS that data delimiter in the split2.csv is the comma.

FIRSTOBS=1 tells SAS to start to read the split2.csv data from row 1.

obs=500 tells SAS to stop reading the split2.csv once row 500 is read.

INPUT lists variables with their INFORMATs. Both the region and StateID are character variables evidenced by the $ INFORMAT, and population is a standard numerical variable.

RUN kicks off the DATA step.

PROC EXPORT invokes SAS's export function. DATA=sd1 tells SAS to export the newly created sd1.sas7bdat file as the sd1.csv file in the folder of c:\sas_exercise specified with the OUTFILE command.

The DBMS (database management systems) command tells SAS that the sd1.csv data is going to have the , comma as the observation delimiter showed by the DMBS=dlm and delimiter=',' commands.

The REPLACE instructs SAS to replace the to-be-created sd1.csv file in the c:\sas_exercise folder if there is already one there.

RUN kicks off the PROC EXPORT procedure.

```
DATA sd2;
INFILE 'c:\sas_exercise\split2.csv' DLM=',' FIRSTOBS=501 obs=800 ;
INPUT ID region $     StateID $ population ;
RUN;
PROC EXPORT DATA=sd2 OUTFILE ='c:\sas_exercise\sd2.csv'
DBMS =dlm REPLACE;
delimiter=',';
RUN;
```

DATA sd2 creates a new SAS dataset named sd2.sas7bdat and stores in the default WORK library.

INFILE tells SAS that the to-be-created new sd2.sas7bdat is based on the external split2.csv data file stored in the c:\sas_exercise folder.

DLM=',' tells SAS that data delimiter in the split2.csv is the comma.
FIRSTOBS=501 tells SAS to start to read the split2.csv data from row 501.
obs=800 tells SAS to stop reading the split2.csv once row 800 is read.
INPUT lists variables with their INFORMATs. Both the region and StateID are character variables evidenced by the $ INFORMAT, and population is a standard numerical variable.
RUN kicks off the DATA step.
PROC EXPORT invokes SAS's export function. DATA=sd2 tells SAS to export the newly created sd2.sas7bdat file as the sd2.csv file in the folder of c:\sas_exercise as specified by the OUTFILE command.
The DBMS (database management systems) command tells SAS that the sd2.csv data is going to have the , comma as the observation delimiter showed by the DMBS=dlm and delimiter=',' commands.
The REPLACE instructs SAS to replace the to-be-created sd2.csv file in the c:\sas_exercise folder if there is already one there.
RUN kicks off the PROC EXPORT procedure.

```
DATA sd3;
INFILE 'c:\sas_exercise\split2.csv' DLM=',' FIRSTOBS=801 obs=1000 ;
INPUT ID region $      StateID $ population ;
RUN;
PROC EXPORT DATA=sd3 OUTFILE='c:\sas_exercise\sd3.csv'
DBMS =dlm REPLACE;
delimiter=',';
RUN;
```

DATA sd3 creates a new SAS dataset named sd3.sas7bdat and stored in the default WORK library.
INFILE tells SAS that the to-be-created new sd3.sas7bdat is based on the external split2.csv data file stored in the c:\sas_exercise folder.
DLM=',' tells SAS that data delimiter in the split2.csv is the comma.
FIRSTOBS=801 tells SAS to start to read the split2.csv data from row 801.
obs=1000 tells SAS to stop reading the split2.csv once row 1000 is read.
INPUT lists variables with their INFORMATs. Both the region and StateID are character variables evidenced by the $ INFORMAT, and population is a standard numerical variable.
RUN kicks off the DATA step.
PROC EXPORT invokes SAS's export function. DATA=sd3 tells SAS to export the newly created sd3.sas7bdat file as the sd3.csv file in the folder of c:\sas_exercise specified with the OUTFILE command.
The DBMS (database management systems) command tells SAS that the sd3.csv data is going to have the , comma as the observation delimiter showed by the DMBS=dlm and delimiter=',' commands.
The REPLACE instructs SAS to replace the to-be-created sd3.csv file in the c:\sas_exercise folder if there is already one there.
RUN kicks off the PROC EXPORT procedure.
The resulting folder shows the three newly created files called sd1.csv, sd2.csv, and sd3.csv.

Result:

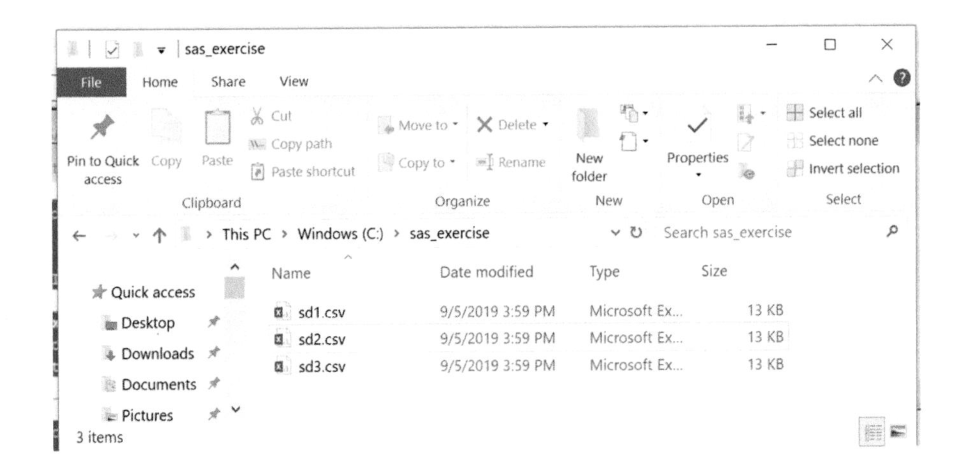

4.2.2 Dividing a SAS Data File Based on Data Attributes

You can use your data attributes to split your dataset. For continuous variables, you can specify ranges to group observations of a variable.

Example: 4.2.2

Data: section422speed.csv
The data file has a total of 65,535 observations (rows) in the format illustrated below.

Obs	ID	Mdate	Month	Mtime	speed	month2
1	A118	01JAN2017	1	01JAN2017	54	January
2	A118	01JAN2017	1	01JAN2017	58	January
3	A118	01JAN2017	1	01JAN2017	57	January
4	A118	01JAN2017	1	01JAN2017	59	January
5	A118	01JAN2017	1	01JAN2017	57	January
6	A118	01JAN2017	1	01JAN2017	61	January
7	A118	01JAN2017	1	01JAN2017	54	January
8	A118	01JAN2017	1	01JAN2017	60	January
9	A118	01JAN2017	1	01JAN2017	60	January
10	A118	01JAN2017	1	01JAN2017	56	January
11	A118	01JAN2017	1	01JAN2017	61	January
12	A118	01JAN2017	1	01JAN2017	62	January
13	A118	01JAN2017	1	01JAN2017	60	January
14	A118	01JAN2017	1	01JAN2017	58	January
15	A118	01JAN2017	1	01JAN2017	60	January
16	A118	01JAN2017	1	01JAN2017	42	January

Task: Split the speed data file into:

1. 12 monthly files (**month2** is a discrete variable)
2. Three-speed files. First one: speed= >55 mph; second one: 35 mph< =speed< 55 mph; and the third one: speed<35 mph (**speed** is a continuous variable).

SAS Codes (E422)

```
LIBNAME ms 'c:\sas_exercise';
DATA ms.jan ms.feb ms.mar ms.apr ms.may ms.jun ms.jul ms.aug
ms.sep ms.oct ms.nov ms.dec ms.hspeed ms.speed1 ms.speed2;
SET ms.sp_speed;
IF Month2='January' THEN OUTPUT ms.jan;
ELSE IF Month2='February' THEN OUTPUT ms.feb;
ELSE IF Month2='March' THEN OUTPUT MS.MAR;
ELSE IF Month2='April' THEN OUTPUT ms.apr;
ELSE IF Month2='May' THEN OUTPUT ms.may;
ELSE IF Month2='June' THEN OUTPUT ms.jun;
ELSE IF Month2='July' THEN OUTPUT ms.jul;
ELSE IF Month2='August'  THEN OUTPUT ms.aug;
ELSE IF Month2='September'  THEN OUTPUT ms.sep;
ELSE IF Month2='October'  THEN OUTPUT ms.oct;
ELSE IF Month2='November'  THEN OUTPUT ms.nov;
ELSE IF Month2='December' THEN OUTPUT ms.dec;
IF speed=>55 THEN OUTPUT ms.Hspeed;
ELSE IF 35<=speed<55 THEN OUTPUT ms.speed1;
ELSE IF Speed<35 THEN OUTPUT ms.speed2;
RUN;
```

LIBNAME defines a library named ms representing the folder of c:\sas_exercise.

DATA creates 15 new SAS data files with the two-layer naming convention. For example, ms.jan stands for the jan.sas7bdat SAS data file to be created and stored in the library of ms (given ms is not the WORK library, the jan.sas7bdat file created is permanent).

SET ms.sp_speed declares to SAS that all the to-be-created 15 files are based on the sp_speed.sas7bdat stored in the ms library (c:\sas_exercise).

The statement of IF Month2="January" THEN OUTPUT ms.jan instructs SAS that if observation for variable Month2 equals to January, then the observation (row) should be directed to the file Jan.sas7bdat in the ms library.

Pay attention to the apostrophe ' ' used to enclose January. A pair of apostrophe or quotation marks must be used for 'January' ("January") because the observation is a character type.

Statements of ELSE IF continue other conditional alternatives until all are listed.

Statements of If speed =>55 THEN OUTPUT … ELSE IF… continue to create new datasets based on different conditions.

RUN kicks off the DATA step.

Results:
15 files were created based on the original file.

Windows (C:) › sas_exercise

Name	Date modified	Type	Size
apr.sas7bdat	10/9/2019 9:17 AM	SAS Data Set	277 KB
aug.sas7bdat	10/9/2019 9:17 AM	SAS Data Set	249 KB
dec.sas7bdat	10/9/2019 9:17 AM	SAS Data Set	249 KB
feb.sas7bdat	10/9/2019 9:17 AM	SAS Data Set	257 KB
hspeed.sas7bdat	10/9/2019 9:17 AM	SAS Data Set	1,453 KB
jan.sas7bdat	10/9/2019 9:17 AM	SAS Data Set	285 KB
jul.sas7bdat	10/9/2019 9:17 AM	SAS Data Set	281 KB
jun.sas7bdat	10/9/2019 9:17 AM	SAS Data Set	277 KB
mar.sas7bdat	10/9/2019 9:17 AM	SAS Data Set	285 KB
may.sas7bdat	10/9/2019 9:17 AM	SAS Data Set	285 KB
nov.sas7bdat	10/9/2019 9:17 AM	SAS Data Set	241 KB
oct.sas7bdat	10/9/2019 9:17 AM	SAS Data Set	249 KB
sep.sas7bdat	10/9/2019 9:17 AM	SAS Data Set	241 KB
speed1.sas7bdat	10/9/2019 9:17 AM	SAS Data Set	1,473 KB
speed2.sas7bdat	10/9/2019 9:17 AM	SAS Data Set	209 KB

4.3 JOINING DATA FILES SEQUENTIALLY (VERTICALLY)

You may encounter the need to join different datasets sequentially in SAS during data analysis. Two basic scenarios you may encounter during the procedure are (a) complete identical columns and (b) partial identical columns. The examples below illustrate how you may handle such situations.

4.3.1 Joining Data Files with Identical Variables (Columns)

Before you try to join any files together, you will need to input all the files into SAS. Datasets listed following the **SET** command need to be of the .sas7bdat type. Other types of data need to be read into SAS with the **INPUT** statement.

Illustration

Data:

originaldata1.sas7bdat,
originaldata2.sas7bdat, and
originaldata3.sas7bdat.

These three files are stored in the folder of c:\sas_exercise.

Task: Join these three files sequentially by creating a new permanent file called newJointData.sas7bdat.

SAS Codes (E431):

```
LIBNAME myjoin
'c:\sas_exercise';
DATA myjoin.newJointData;
SET  myjoin.OriginalData1
myjoin.OriginalData2
myjoin.OriginalData3;
RUN;
```

LIBNAME myjoin "c:\sas_exercise" defines a library named myjoin representing the directory of c:\sas_exercise.

DATA myjoin.newJointData creates a new permanent dataset called newJointData.sas7bdat and stores it in the myjoin library.

SET statement lists all data files in the myjoin library to be joined vertically. Here OriginalData1, OriginalData2, and OriginalData3 are to be joined together vertically.

RUN kicks off the DATA step.

Example: 4.3.1

Data: WSC.csv and WNC.csv.
WSC.csv

Obs	region	StateID	population
1	WSCentral	AR	3013825
2	WSCentral	LA	4659978
3	WSCentral	OK	3943079
4	WSCentral	TX	28701845

WNC.csv

Obs	region	StateID	population
1	WNCentral	IA	3156145
2	WNCentral	KS	2911505
3	WNCentral	MN	5611179
4	WNCentral	MO	6216452
5	WNCentral	ND	760077
6	WNCentral	NE	1929268
7	WNCentral	SD	882235

Task: Join the two files together in sequential order (vertically).

SAS Codes (E431a) – generating a temporary file
The code below merges the 2 files and produces a file located in the default **WORK** library.

```
DATA wsc;
INFILE 'c:\sas_exercise\wsc.csv' DLM=',' DSD FIRSTOBS=2 ;
```

```
INPUT region :$9.  StateID $ population ;
RUN;
DATA wnc;
INFILE 'c:\sas_exercise\wnc.csv' DLM=',' DSD FIRSTOBS=2 ;
INPUT region :$9.  StateID $ population ;
RUN;
DATA wsc_wnc;
   SET wsc wnc;
RUN;
PROC PRINT DATA=wsc_wnc;
RUN;
```

DATA wsc creates a new temporary SAS dataset called wsc.sas7bdat in the WORK library.

INFILE "c:\sas_exercise\wsc.csv" specifies the directory path and the file wsc.csv to be read and used to create the wsc.sas7bdat file. DLM=',' tells SAS that the delimiter used in the wsc.csv file is a comma.

DSD further instructs SAS to handle three things:

1: Treat two consecutive delimiters (, ,) as a missing value.

2: Remove quotation marks from character values and treat everything within a quotation mark together as a single observation for the variable (e.g., "AB, C", XYZ will be treated as 2 variables, not 3. The "AB, C" is treated as 1).

3: Assume data separator is the , comma.

FIRSTOBS=2 tells SAS to start to read data from row 2.

INPUT region :$9. StateID $ population instructs SAS to (a) read up to 9 columns for the character variable region (colon: means up to the nth columns or until a separator is encountered) and (b) treat StateID as a character variable and population as a numerical variable.

RUN kicks off the DATA step.

DATA wnc creates a new temporary SAS dataset called wnc.sas7bdat in the WORK library.

INFILE "c:\sas_exercise\wnc.csv" specifies the directory path and the file wnc.csv to be read and used to create the wnc.sas7bdat file. DLM=',' tells SAS that the delimiter used in the wnc.csv file is the comma.

FIRSTOBS=2 tells SAS that the SAS should start to read data from row 2.

INPUT region :$9. StateID $ population instructs SAS to (a) read up to 9 columns for the character variable region (colon: means up to the nth columns or until a separator is encountered) and (b) treat StateID as a character variable and population as a numerical variable.

RUN kicks off the DATA step.

DATA wsc_wnc creates a new temporary SAS dataset called wsc_wnc.sas7bdat in the WORK library.

SET wsc wnc instructs SAS to use the 2 just created temporary files named wsc.sas7bdat and wnc.sas7bdat for the to-be-created new dataset wsc-wnc.sas7bdat.

RUN kicks off this DATA step.

PROC PRINT DATA invokes the print procedure to print the data file named wsc_wnc.sas7bdat.

RUN kicks off this PROC PRINT step.

Result: The wsc_wnc.sas7bdat resides in the WORK library.

Obs	region	StateID	population
1	WSCentral	AR	3013825
2	WSCentral	LA	4659978
3	WSCentral	OK	3943079
4	WSCentral	TX	28701845
5	WNCentral	IA	3156145
6	WNCentral	KS	2911505
7	WNCentral	MN	5611179
8	WNCentral	MO	6216452
9	WNCentral	ND	760077
10	WNCentral	NE	1929268
11	WNCentral	SD	882235

SAS Codes (E431b) – generating a permanent file
The code below merges the two files and generates an external wsc_wnc.sas7bdat data file in the folder c:\sas_exercise.

```
DATA wsc;
INFILE 'c:\sas_exercise\wsc.csv' DLM=',' DSD FIRSTOBS=2 ;
INPUT region :$9.  StateID $ population ;
RUN;
DATA wnc;
INFILE 'c:\sas_exercise\wnc.csv' DLM=',' DSD FIRSTOBS=2 ;
INPUT region :$9.  StateID $ population ;
RUN;
DATA wsc_wnc;
   SET wsc wnc;
RUN;

LIBNAME mymerge 'c:\sas_exercise';
DATA mymerge.wsc_wnc;
   SET wsc wnc;
RUN;
```

Continuing from the explanation of SAS Codes E431a

LIBNAME mymerge "c:\sas_exercise" defines a library called mymerge representing the folder of c:\sas_exercise.

DATA mymerge.wsc_wnc creates a new permanent SAS dataset called wsc_wnc.sas7bdat and stores it in the mymerge library.

SET wsc wnc instructs SAS to use the 2 just created temporary files named wsc.sas7bdat and wnc.sas7bdat for the to-be-created new dataset wsc_wnc.sas7bdat.

RUN kicks off this DATA step.

Result:

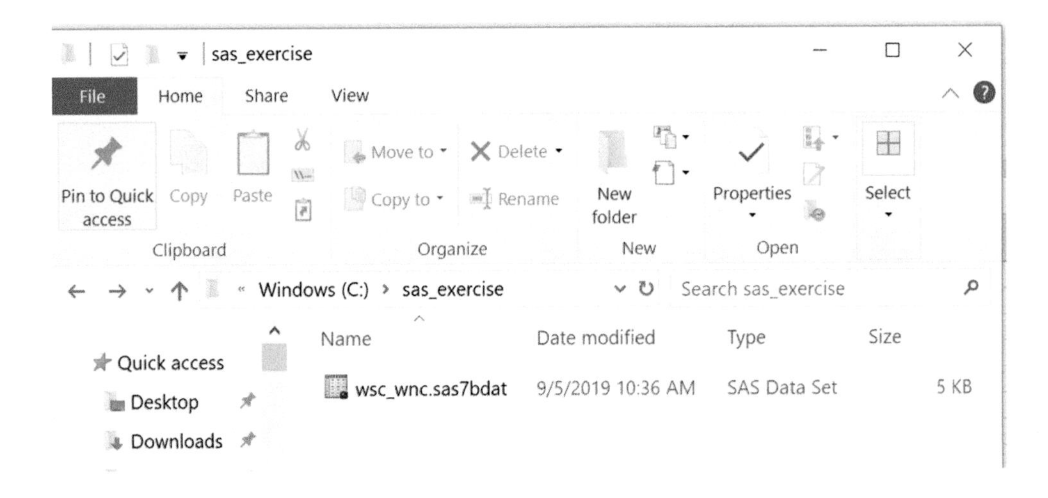

SAS Codes (E431c) - generating a csv file

The code below merges the two files and generates an external wsc_wnc.csv file in the folder c:\sas_exercise.

```
DATA wsc;
INFILE 'c:\sas_exercise\wsc.csv' DLM=',' DSD FIRSTOBS=2 ;
INPUT region :$9.  StateID $ population ;
RUN;
DATA wnc;
INFILE 'c:\sas_exercise\wnc.csv' DLM=',' DSD FIRSTOBS=2 ;
INPUT region :$9.  StateID $ population ;
RUN;
DATA wsc_wnc;
   SET wsc wnc;
RUN;
LIBNAME mymerge 'c:\sas_exercise';
DATA mymerge.wsc_wnc;
   SET wsc wnc;
RUN;
PROC EXPORT DATA=mymerge.wsc_wnc OUTFILE='c:\sas_exercise\wsc_wnc.
csv'
DBMS =dlm REPLACE;
Delimiter=','  ;
RUN;
```

DATA wsc creates a new temporary SAS dataset called wsc.sas7bdat in the WORK library.

INFILE "c:\sas_exercise\wsc.csv" specifies the directory path and the file wsc.csv to be read and used to create the wsc.sas7bdat file. DLM=',' tells SAS that the delimiter used in the wsc.csv file is a comma.

DSD further instructs SAS to handle three things:

1: Treat two consecutive delimiters (, ,) as a missing value.

2: Remove quotation marks from character values and treat everything within a quotation mark together as a single observation for the variable (e.g., "AB, C", XYZ will be treated as 2 variables, not 3. The "AB, C" is treated as 1).

3: Assume data separator is the , comma.

FIRSTOBS=2 tells SAS to read data from row 2.

INPUT region :$9. StateID $ population instructs SAS to (a) read up to 9 columns for the character variable region (colon: means up to the nth columns or until a separator is encountered) and (b) treat StateID as a character variable and population as a numerical variable.

RUN kicks off the DATA step.

DATA wnc creates a new temporary SAS dataset called wnc.sas7bdat in the WORK library.

INFILE "c:\sas_exercise\wnc.csv" specifies the directory path and the file wnc.csv to be read and used to create the wnc.sas7bdat file. DLM=',' tells SAS that the delimiter used in the wnc.csv file is the comma.

FIRSTOBS=2 tells SAS to read data from row 2.

INPUT region :$9. StateID $ population instructs SAS to (a) read up to 9 columns for the character variable region (colon: means up to the nth columns or until a separator is encountered) and (b) treat StateID as a character variable and population as a numerical variable.

RUN kicks off the DATA step.

DATA wsc_wnc creates a new temporary SAS dataset called wsc_wnc.sas7bdat in the WORK library.

SET wsc wnc instructs SAS to use the 2 just created temporary files named wsc.sas7bdat and wnc.sas7bdat for the to-be-created new dataset wsc_wnc.sas7bdat.

RUN kicks off this DATA step.

LIBNAME mymerge "c:\sas_exercise" defines a library called mymerge representing the folder of c:\sas_exercise.

DATA mymerge.wsc_wnc creates a new permanent SAS dataset called wsc_wnc.sas7bdat in the mymerge library.

SET wsc wnc instructs SAS to use the 2 just created temporary files named wsc.sas7bdat and wnc.sas7bdat for the to-be-created new dataset.

RUN kicks off this DATA step.

PROC EXPORT invokes the export procedure. DATA=mymerge.wsc_wnc tells SAS to export the wsc_wnc.sas7bdat file in the mymerge library to an external file called wsc_wnc.csv in the folder of c:\sas_exercise with the OUTFILE command.

DBMS=dlm tells SAS that the wsc_wnc.csv will have a delimiter to separate values and the delimiter is the comma (delimitier=','). REPLACE tells SAS to override any other file in the directory, which has the same name as wsc_wnc.csv.

RUN kicks off this PROC EXPORT step.

Result:

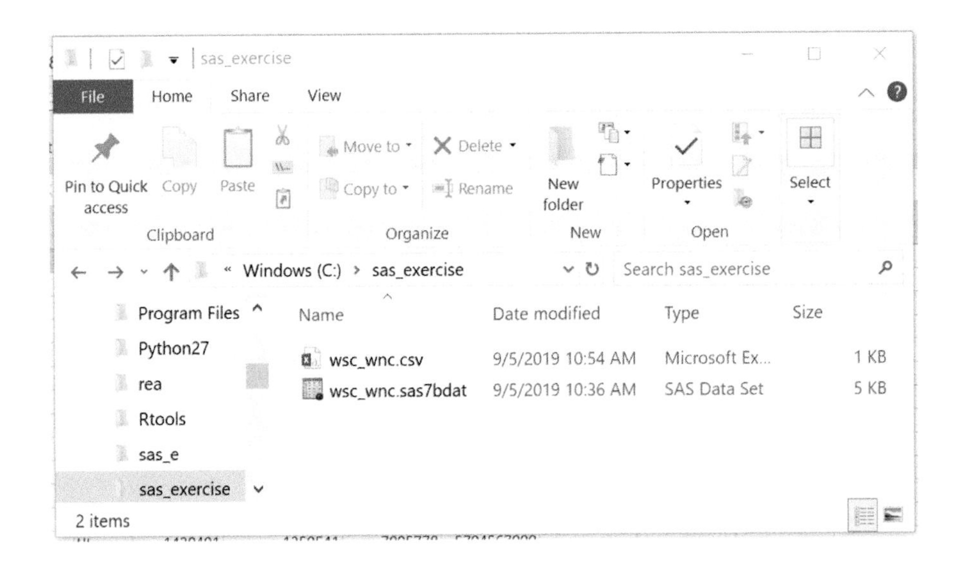

4.3.2 Joining Files Having Different Variables (Columns)

More often than not, with sequential file merging, you will encounter files that have different variables. While the merging procedure is the same as merging identical variables, you should be aware that the newly merged file possesses all variables contained in all files. And you do not need to sort data in any order before joining.

Example: 4.3.2

Data:
wsc.csv data

Obs	region	StateID	population
1	WNCentral	IA	3156145
2	WNCentral	KS	2911505
3	WNCentral	MN	5611179
4	WNCentral	MO	6216452
5	WNCentral	ND	760077
6	WNCentral	NE	1929268
7	WNCentral	SD	882235

NewEngland.csv

Obs	region	StateID	vehicles	population
1	NewEngland	CT	2826350	3572665
2	NewEngland	MA	5065221	6902149
3	NewEngland	ME	1094387	1338404
4	NewEngland	NH	1319117	1356458
5	NewEngland	RI	872814	1057315
6	NewEngland	VT	621608	626299

Task: Merge the wsc.csv and the newengland.csv

SAS Codes (E432)

```
DATA wsc;
FORMAT region $10.;
INFILE 'c:\sas_exercise\wsc.csv' DLM=',' FIRSTOBS=2 ;
INPUT region :$9.  StateID $ population ;
RUN;
DATA newengland;
INFILE 'c:\sas_exercise\newengland.csv' DLM=',' FIRSTOBS=2 ;
INPUT region :$10.  StateID $ vehicles population ;
RUN;
LIBNAME mymerge 'c:\sas_exercise';
DATA mymerge.wsc_newengland;
   SET wsc newengland;
RUN;
PROC EXPORT DATA=mymerge.wsc_newengland
OUTFILE='c:\sas_exercise\wsc_newengland.csv'
DBMS =dlm REPLACE;
delimiter=',';
RUN;
```

DATA wsc creates a new temporary SAS dataset called wsc.sas7bdat in the WORK library.

FORMAT region $10. tells SAS to display the character variable region with 10-column width.

INFILE "c:\sas_exercise\wsc.csv" specifies the directory path for the file wsc.csv to be read and used to create the wsc.sas7bdat file. DLM=',' tells SAS that the delimiter used in the wsc.csv file is the , comma.

FIRSTOBS=2 tells SAS to read data from row 2.

INPUT region :$9. StateID $ population instructs SAS to (a) read up to 9 columns for the character variable region (colon: means up to the nth columns or until a separator is encountered) and (b) treat StateID as a character variable and population as a numerical variable.

RUN kicks off the DATA step.

DATA newengland creates a new temporary SAS dataset called newengland.sas7bdat in the WORK library.

INFILE "c:\sas_exercise\newengland.csv" specifies the directory path and the file newengland.csv to be read and used to create the newengland.sas7bdat file. DLM=',' tells SAS that the delimiter used in the newengland.csv file is the , comma.

FIRSTOBS=2 tells SAS that the SAS should start to read data from row 2.

INPUT region :$10. StateID $ vehicles population instructs SAS to (a) read up to 10 columns for the character variable region (colon: means up to the nth columns or until a separator is encountered) and (b) treat StateID as a character variable, and vehicles and population as numerical variables.

RUN kicks off the DATA step.

LIBNAME mymerge "c:\sas_exercise" defines a library called mymerge representing the folder of c:\sas_exercise.

DATA mymerge.wsc_newengland creates a new permanent SAS dataset called wsc_newengland.sas7bdat in the mymerge library.

SET wsc newengland instructs SAS to use the 2 just created temporary files named wsc.sas7bdat and newengland.sas7bdat for the to-be-created new dataset.

RUN kicks off this DATA step.

PROC EXPORT invokes the export procedure. DATA=mymerge.wsc_newengland tells SAS to export the wsc_newengland.sas7bdat file in the mymerge library to an external file called wsc_newengland.csv in the folder of c:\sas_exercise with the OUTFILE command.

DBMS=dlm tells SAS that the wsc_newengland.csv will have a delimiter to separate values and the delimiter is the comma (delimitier=','). REPLACE tells SAS to override any other file in the directory, which has the same name as wsc_newengland.csv.

RUN kicks off this PROC EXPORT procedure.

Result:

Obs	region	StateID	population	vehicles
1	WSCentral	AR	3013825	.
2	WSCentral	LA	4659978	.
3	WSCentral	OK	3943079	.
4	WSCentral	TX	28701845	.
5	NewEngland	CT	3572665	2826350
6	NewEngland	MA	6902149	5065221
7	NewEngland	ME	1338404	1094387
8	NewEngland	NH	1356458	1319117
9	NewEngland	RI	1057315	872814
10	NewEngland	VT	626299	621608

Result:

All columns (variables) are shown

There are four missing vehicle observations. This is because the wsc.csv dataset only contains StateID and population variables.

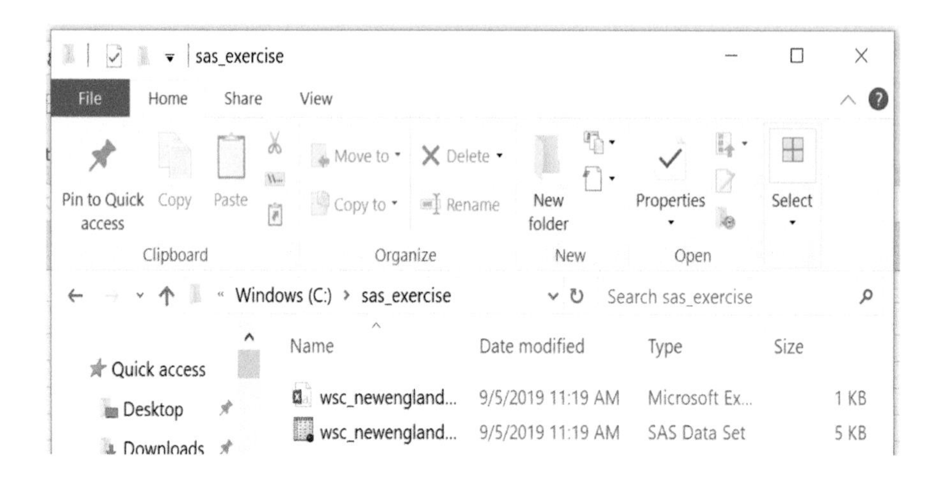

4.4 JOINING FILES HORIZONTALLY (LEFT/RIGHT)

All relational databases can be linked horizontally through a common unique data variable (the unique ID) for efficient processing.

A horizontal (left/right) join can be a one-to-one join (one row links to exactly one other row) or one-to-many join (one row links to many other rows). During data analysis, joining files horizontally is a fairly regular practice.

4.4.1 One-to-One Merge

In a one-to-one merge, the two files being merged have a one-to-one by row relationship – meaning rows are matched on a one-to-one basis based on a shared common data variable.

Example: 4.4.1

Data:

Pop2.sas7bdat (See actual data on next page)
dri2.csv

The variable State (column) in both datasets is unique. The two datasets have a one-to-one relationship based on the State variable.

The SAS System

Obs	region	State	Drivers
1	Pacific	AK	4329503
2	ESCentral	AL	5377653
3	WSCentral	AR	534585
4	Mountain	AZ	6301363
5	Pacific	CA	1032703
6	Mountain	CO	1473262
7	NewEnglan	CT	2053924
8	Southatla	DC	951008
9	Southatla	DE	1190367
10	Southatla	FL	8529404
11	Southatla	GA	4553584
12	Pacific	HI	7095778
13	WNCentral	IA	26777132
14	Mountain	ID	1918305
15	ENCentral	IL	5768281

Obs	region	state	populations
1	Pacific	AK	737438
2	ESCentral	AL	4887871
3	WSCentral	AR	3013825
4	Mountain	AZ	7171646
5	Pacific	CA	39557045
6	Mountain	CO	5695564
7	NewEngland	CT	3572665
8	Southatlantic	DC	702455
9	Southatlantic	DE	967171
10	Southatlantic	FL	21299325
11	Southatlantic	GA	10519475
12	Pacific	HI	1420491
13	WNCentral	IA	3156145
14	Mountain	ID	1754208
15	ENCentral	IL	12741080

Task: Merge the two files

SAS Codes (E441)

```
LIBNAME mydat 'c:\sas_exercise';
DATA Newpop;
SET mydat.pop2;
RUN;
```

```
PROC SORT DATA=Newpop;
BY State;
PROC PRINT DATA=Newpop;
RUN;

DATA Newdri;
INFILE 'c:\sas_exercise\dri2.csv' DLM=',' FIRSTOBS=2 DSD MISSOVER;
INPUT region :$9. State $ Drivers;
RUN;
PROC SORT DATA=Newdri;
BY State;
PROC PRINT DATA=Newdri;
RUN;

DATA mydat.pop_dri;
MERGE      Newpop   Newdri;
BY state;
RUN;

PROC PRINT DATA=mydat.pop_dri;
RUN;
```

LIBNAME mydat "c:\sas_exercise" defines a library called mydat representing the folder of c:\sas_exercise.

DATA Newpop creates a new temporary SAS dataset called Newpop.sas7bdat in the WORK library.

SET mydat.pop2 instructs SAS to use a file named pop2.sas7bdat in the library of mydat for the to-be-created new dataset.

RUN kicks off this DATA step.

PROC SORT invokes the sort procedure. DATA=Newpop; BY State tells SAS to sort the variable State in the Newpop.sas7bdat file in the temporary WORK library in an ascending order (A to Z).

PROC PRINT DATA=Newpop invokes the print procedure to print the entire Newpop.sas7bdat on the screen.

RUN kicks off the PROC PRINT procedure.

DATA Newdri creates a new temporary SAS dataset called Newdri.sas7bdat in the WORK library.

INFILE "c:\sas_exercise\dri2.csv" specifies the directory path and the file dri2.csv to be read and used to create the Newdri.sas7bdat file.

DLM=',' tells SAS that the delimiter used in the dri2.csv file is , comma.

FIRSTOBS=2 tells SAS to read data from row 2.

DSD further instructs SAS to handle three things:

1: Treat two consecutive delimiters (, ,) as a missing value.

2: Remove quotation marks from character values and treat everything within a quotation mark together as a single observation for the variable (e.g., "AB, C", XYZ will be treated as 2 variables, not 3. The "AB, C" is treated as 1).

3: Assume data separator is the , comma.

MISSOVER prevents SAS from going to the next data line if it can't locate values with the current dateline for all variables listed in the INPUT statement. SAS assigns a missing value for variables without values.

INPUT region :$9. State $ Drivers instructs SAS to (a) read up to 9 columns for the character variable region (colon: means up to the nth columns or until a separator is encountered) and (b) treat State as a character variable and Drivers as a numerical variable.

RUN kicks off the DATA step.

PROC SORT invokes the sort procedure. DATA=Newdri; BY State tells SAS to sort the variable State in the Newdri.sas7bdat file in the temporary WORK library in an ascending order (A to Z).

PROC PRINT DATA=Newdri invokes the print procedure to print the entire Newdri.sas7bdat on the screen.

RUN kicks off the PROC PRINT procedure.

DATA mydat.pop_dri creates a permanent pop_dri.sas7bdat in the mydat library.

The MERGE command instructs SAS to merge the just created Newpop and Newdri based on state as specified by the BY state statement.

RUN kicks off the DATA step.

PROC PRINT DATA=mydat.pop_dri outputs the file pop_dri.sas7bdat in the mydat library on the screen.

RUN kicks off the PROC PRINT procedure.

Result:

Obs	region	state	populations	Drivers
1	Pacific	AK	737438	4329503
2	ESCentral	AL	4887871	5377653
3	WSCentral	AR	3013825	534585
4	Mountain	AZ	7171646	6301363
5	Pacific	CA	39557045	1032703
6	Mountain	CO	5695564	1473262
7	NewEnglan	CT	3572665	2053924
8	Southatla	DC	702455	951008
9	Southatla	DE	967171	1190367
10	Southatla	FL	21299325	8529404
11	Southatla	GA	10519475	4553584
12	Pacific	HI	1420491	7095778
13	WNCentral	IA	3156145	26777132
14	Mountain	ID	1754208	1918305
15	ENCentral	IL	12741080	5768281
16	ENCentral	IN	6691878	560247
17	WNCentral	KS	2911505	4156138
18	ESCentral	KY	4468402	17099340
19	WSCentral	LA	4659978	3954378

4.4.2 One-to-Many Join

With a one-to-many join (merge), the two files to be merged have a different row number. See the example below.

Example: 4.4.2

Data: county_inf.csv and rc.csv
County_inf.csv

Obs	county	state	Sfips	pop
1	Apache	AZ	4	.
2	Gila	AZ	4	.
3	Pinal	AZ	4	.
4	Fairfield	CT	9	916829
5	Hartford	CT	9	894014
6	Litchfiel	CT	9	189927
7	Middlesse	CT	9	165676
8	NewHaven	CT	9	862477
9	NewLondon	CT	9	274055
10	Tolland	CT	9	152691
11	Windham	CT	9	118428
12	Bristol	RI	44	49785
13	Kent	RI	44	166325

Region Data: rc.csv

Obs	region	State
1	NewEngland	CT
2	WNCentral	NE
3	NewEngland	RI
4	WSCentral	TX

Task: Merge the two files so every county will also have its corresponding region information.

SAS Codes (E442)

```
LIBNAME mydat 'c:\sas_exercise';
DATA region2;
FORMAT region $13.;
INFILE 'c:\sas_exercise\rc.csv' DLM=',' FIRSTOBS=2 DSD MISSOVER;
INPUT region $          State $;
RUN;
```

```
PROC SORT DATA=region2;
BY State;
RUN;

DATA county2;
FORMAT county $9.;
INFILE 'C:\sas_exercise\county_inf.csv' DLM=','  FIRSTOBS=2 DSD
MISSOVER;
INPUT state $    Sfips    County $    pop;
RUN;

PROC SORT DATA=county2;
BY State;

DATA mydat.region_county;
MERGE      region2 county2;
BY state;
RUN;

DATA mydat.county_region;
MERGE      county2   region2;
BY state;
RUN;

PROC PRINT DATA=mydat.county_region;
RUN;
```

LIBNAME mydat "c:\sas_exercise" defines a library called mydat representing the folder of c:\sas_exercise.

DATA region2 creates a new temporary SAS dataset called region2.sas7bdat in the WORK library.

FORMAT region $13. instructs SAS to display the character variable region with 13-column width.

INFILE "c:\sas_exercise\rc.csv" specifies the directory path and the file rc.csv to be read and used to create the region2.sas7bdat file.

DLM=',' tells SAS that the delimiter used in the rc.csv file is , comma.

FIRSTOBS=2 tells SAS that the SAS should start to read data from row 2.

DSD further instructs SAS to handle three things:

1: Treat two consecutive delimiters (, ,) as a missing value.

2: Remove quotation marks from character values and treat everything within a quotation mark together as a single observation for the variable (e.g., "AB, C", XYZ will be treated as 2 variables, not 3. The "AB, C" is treated as 1).

3: Assume data separator is the , comma.

MISSOVER prevents SAS from going to the next data line if it can't locate values with the current dateline for all variables listed in the INPUT statement. SAS assigns a missing value for variables without values.

INPUT region $ State $ tells SAS that the observations to be read are for the character variables of region and State.

RUN kicks off the DATA step.

PROC SORT invokes the sort procedure. DATA=region2; BY State tells SAS to sort the variable State in the region2.sas7bdat file in the temporary WORK library in an ascending order (A to Z).

RUN kicks off the PROC SORT procedure.

DATA county2 creates a new temporary SAS dataset called county2.sas7bdat in the WORK library.

FORMAT county $9. instructs SAS to display the character variable county with 9-column width.

INFILE "c:\sas_exercise\county_inf.csv" specifies the directory path and the file county_inf.csv to be read and used to create the county2.sas7bdat file.

DLM=',' tells SAS that the delimiter used in the county_inf.csv file is , comma.

FIRSTOBS=2 tells SAS that the SAS should start to read data from row 2.

DSD further instructs SAS to handle three things:

1: Treat two consecutive delimiters (, ,) as a missing value.

2: Remove quotation marks from character values and treat everything within a quotation mark together as a single observation for the variable (e.g., "AB, C", XYZ will be treated as 2 variables, not 3. The "AB, C" is treated as 1).

3: Assume data separator is the , comma.

MISSOVER prevents SAS from going to the next data line if it can't locate values with the current dateline for all variables listed in the INPUT statement. SAS assigns a missing value for variables without values.

INPUT state $ Sfips County $ pop instructs SAS to read observations in the order of state, sfips, County, and pop where state and County are character variables, and sfips and pop are standard numerical variables.

RUN kicks off the DATA step.

PROC SORT invokes the sort procedure. DATA=county2; BY State tells SAS to sort the variable state in the county2.sas7bdat file located at the temporary WORK library in an ascending order (A to Z).

DATA mydat.region_county creates a permanent region_county.sas7bdat file in the mydat library.

The MERGE region2 county2 BY state statement instructs SAS to merge the county2 and region2 datasets based on matching of state for the creation of region_county.sas7bdat dataset.

RUN kicks off the DATA step.

PROC PRINT DATA=mydat.region_county outputs the file region_county.sas7bdat in the mydat library on the screen.

Run kicks off the PROC PRINT procedure.

Results:

The SAS System

Obs	county	state	Sfips	pop	region
1	Apache	AZ	4	.	
2	Gila	AZ	4	.	
3	Pinal	AZ	4	.	
4	Fairfield	CT	9	916829	NewEngland
5	Hartford	CT	9	894014	NewEngland
6	Litchfiel	CT	9	189927	NewEngland
7	Middlesse	CT	9	165676	NewEngland
8	NewHaven	CT	9	862477	NewEngland
9	NewLondon	CT	9	274055	NewEngland
10	Tolland	CT	9	152691	NewEngland
11	Windham	CT	9	118428	NewEngland
12		NE	.	.	WNCentral
13	Bristol	RI	44	49785	NewEngland
14	Kent	RI	44	166325	NewEngland
15	Newport	RI	44	82999	NewEngland
16	Providence	RI	44	629668	NewEngland
17	Washington	RI	44	129125	NewEngland
18		TX	.	.	WSCentral

4.4.3 Tracking Source of Data when Merging by Using the "IN=" Statement

When the **"IN="** statement is used with the **MERGE** statement, it generates a new temporary variable for each of the to-be-merged datasets with a value of either 1 or 0. This temporary variable keeps track of where rows in the merged dataset are coming from (which dataset contributes to which row in the merged dataset).

The "IN=" variable for a dataset is automatically assigned 1 when the dataset contributes to a row of the merged file. The "IN=" for a dataset is assigned 0 when the dataset does not contribute to the merged file. See the example below for a complete understanding of these logical operations.

Example: 4.4.3

Data:

Region data

Region data

Region	State
NewEngland	CT
WNCentral	NE
NewEngland	RI
WSCentral	TX

County data

County Data

Obs	county	state	Sfips	pop
1	Apache	AZ	4	.
2	Gila	AZ	4	.
3	Pinal	AZ	4	.
4	Fairfield	CT	9	916829
5	Hartford	CT	9	894014
6	Litchfiel	CT	9	189927
7	Middlesse	CT	9	165676
8	NewHaven	CT	9	862477
9	NewLondon	CT	9	274055
10	Tolland	CT	9	152691
11	Windham	CT	9	118428
12	Bristol	RI	44	49785
13	Kent	RI	44	166325
14	Newport	RI	44	82999
15	Providence	RI	44	629668
16	Washington	RI	44	129125

Final merged by **state** data:

Obs	county	state	Sfips	pop	IN=	Region	State	IN=
1	Apache	AZ	4	.	1			0
2	Gila	AZ	4	.	1			0
3	Pinal	AZ	4	.	1			0
4	Fairfield	CT	9	916829	1	NewEngland	CT	1
5	Hartford	CT	9	894014	1	NewEngland	CT	1
6	Litchfiel	CT	9	189927	1	NewEngland	CT	1
7	Middlesse	CT	9	165676	1	NewEngland	CT	1
8	NewHaven	CT	9	862477	1	NewEngland	CT	1
9	NewLondon	CT	9	274055	1	NewEngland	CT	1
10	Tolland	CT	9	152691	1	NewEngland	CT	1
11	Windham	CT	9	118428	1	NewEngland	CT	1
12	Bristol	RI	44	49785	1	NewEngland	RI	1
13	Kent	RI	44	166325	1	NewEngland	RI	1
14	Newport	RI	44	82999	1	NewEngland	RI	1
15	Providence	RI	44	629668	1	NewEngland	RI	1
16	Washington	RI	44	129125	1	NewEngland	RI	1
17					0	WNCentral	NE	1
18					0	WSCentral	TX	1

Data: county2.sas7bdat and region2.sas7bdat

Task: Create three subsets of data through merging by **state** of the two datasets discussed above and containing:

a. An inclusive dataset containing all rows
b. A dataset where rows are present in both the original datasets
c. A dataset where rows are only present in one of the original datasets.

SAS Codes (E443)

```
LIBNAME mydat
'c:\sas_exercise';

DATA
mydat.c_r_both_all;
MERGE      mydat.
county2 (IN=c2)
mydat.region2 (IN=r2);
countydata=c2;
regiondata=r2;
BY state;
RUN;
PROC PRINT
DATA=mydat.c_r_both_
all;
RUN;
```

LIBNAME mydat "c:\sas_exercise" defines a library called mydat representing the file path of c:\sas_exercise.

DATA mydat.c_r_both_all creates a new SAS dataset called c_r_both_all.sas7bdat and stores it the library of mydat.

Statements of MERGE mydat.county2 (IN=c2) mydat.region2 (IN=r2); BY state instruct SAS to merge county2 and region2, assign the new temporary variable with a name of c2 through IN=c2 for the county2 dataset and r2 through IN=r2 for region2 dataset (keep in mind that the IN=c2 and IN=r2 temporary variables are automatically generated because the MERGE statement is used).

Since the c2 and r2 are temporary, by using the assignment statements countydata=c2 and regiondata=r2, two new permanent variables countydata and regiondata are created based on the c2 and r2 values.

RUN kicks off the end of the DATA step.

PROC PRINT DATA=mydat.c_r_both_all prints the data called c_r_both_all.sas7bdat stored in the mydat library on the screen.

RUN kicks off the PROC PRINT step.

```
DATA mydat.c_r_both_1;
MERGE      mydat.county2 (IN=c2)  mydat.region2 (IN=r2);
countydata=c2;
regiondata=r2;
BY state;
IF c2=1 and r2=1;
RUN;
PROC PRINT DATA=mydat.c_r_both_1;

RUN;
```

DATA mydat.c_r_both_1 creates a new SAS dataset called c_r_both_1.sas7bdat stored in the library of mydat.

MERGE mydat.county2 (IN=c2) mydat.region2 (IN=r2) instructs SAS to merge county2 and region2. BY state and assigns the new temporary variable c2 through IN=c2 for the county2 dataset and r2 through IN=r2 for region2 dataset (keep in mind that the IN=c2 and IN=r2 temporary variables are automatically generated because the MERGE statement is used).

Since c2 and r2 are temporary, by using the assignment statements countydata=c2 and regiondata=r2, two new permanent variables countydata and regiondata are created based on the IN=c2 and IN=r2 in the c_r_both_1.sas7bdat dataset.

Observations to be kept in the c_r_both_1.sas7bdat are those that meet the condition of c2=1 and r2=1, meaning observations for a row are present in both the original county2 and region2 datasets.

RUN kicks off the end of the DATA step.

PROC PRINT DATA=mydat.c_r_both_1 prints the data called c_r_both_1.sas7bdat stored in the mydat library on the screen.

RUN kicks off the PROC PRINT step.

```
DATA mydat.c_r_either_0;
MERGE      mydat.county2 (IN=c2)  mydat.region2 (IN=r2);
countydata=c2;
regiondata=r2;
BY state;
IF c2=0 or r2=0;
RUN;
PROC PRINT DATA=mydat.c_r_either_0;

RUN;
```

DATA mydat.c_r_either_0 creates a new SAS dataset called c_r_either_0.sas7bdat stored in the library of mydat.

MERGE mydat.county2 (IN=c2) mydat.region2 (IN=r2) instructs SAS to merge county2 and region2. BY state assigns the new temporary variable c2 through IN=c2 for the county2 dataset and r2 through IN=r2 for region2 dataset (keep in mind that the IN=c2 and IN=r2 temporary variables are automatically generated because the MERGE statement is used).

Since c2 and r2 are temporary, by using the assignment statements countydata=c2 and regiondata=r2, two new permanent variables countydata and regiondata are created based on the IN=c2 and IN=r2 in the c_r_either_0.sas7bdat dataset.

Observations to be kept in the c_r_either_0.sas7bdat are those that meet the condition of either c2=0 or r2=0, meaning observations for a row are present in only one of the original county2 and region2 datasets.

RUN kicks off the end of the DATA step.

PROC PRINT DATA=mydat.c_r_either_0 prints the data called c_r_either_0.sas7bdat stored in the mydat library on the screen.

RUN kicks off the PROC PRINT step.

Results:

a. All rows – Task a

Obs	state	Sfips	County	pop	region	countydata	regiondata
1	AZ	4	Apache	.		1	0
2	AZ	4	Gila	.		1	0
3	AZ	4	Pinal	.		1	0
4	CT	9	Fairfield	916829	NewEngland	1	1
5	CT	9	Hartford	894014	NewEngland	1	1
6	CT	9	Litchfiel	189927	NewEngland	1	1
7	CT	9	Middlesse	165676	NewEngland	1	1
8	CT	9	NewHaven	862477	NewEngland	1	1
9	CT	9	NewLondon	274055	NewEngland	1	1
10	CT	9	Tolland	152691	NewEngland	1	1
11	CT	9	Windham	118428	NewEngland	1	1
12	NE	.		.	WNCentral	0	1
13	RI	44	Bristol	49785	NewEngland	1	1
14	RI	44	Kent	166325	NewEngland	1	1
15	RI	44	Newport	82999	NewEngland	1	1
16	RI	44	Providence	629668	NewEngland	1	1
17	RI	44	Washington	129125	NewEngland	1	1
18	TX	.		.	WSCentral	0	1

b. Only rows existed in both datasets – task b

The SAS System

Obs	state	Sfips	County	pop	region	countydata	regiondata
1	CT	9	Fairfield	916829	NewEngland	1	1
2	CT	9	Hartford	894014	NewEngland	1	1
3	CT	9	Litchfiel	189927	NewEngland	1	1
4	CT	9	Middlesse	165676	NewEngland	1	1
5	CT	9	NewHaven	862477	NewEngland	1	1
6	CT	9	NewLondon	274055	NewEngland	1	1
7	CT	9	Tolland	152691	NewEngland	1	1
8	CT	9	Windham	118428	NewEngland	1	1
9	RI	44	Bristol	49785	NewEngland	1	1
10	RI	44	Kent	166325	NewEngland	1	1
11	RI	44	Newport	82999	NewEngland	1	1
12	RI	44	Providenc	629668	NewEngland	1	1
13	RI	44	Washingto	129125	NewEngland	1	1

c. Rows only existed in one of the datasets – task c

The SAS System

Obs	state	Sfips	County	pop	region	countydata	regiondata
1	AZ	4	Apache	.		1	0
2	AZ	4	Gila	.		1	0
3	AZ	4	Pinal	.		1	0
4	NE	.		.	WNCentral	0	1
5	TX	.		.	WSCentral	0	1

4.5 VARIABLES – DROP, KEEP, AND RENAME

Unsurprisingly, the fewer variables you have in your dataset, the more efficiently your SAS program will run. When you create a new dataset from an existing data file, you may not want to keep all of the original variables. You can use the DROP, KEEP, and RENAME statements to customize variables associated with your new datasets.

Example: 4.5

Data:

olddata.sas7bdat
olddata.csv

Task: Create a new dataset with only some of the original variables

SAS Codes (E45)

Examples below are using the olddat.sas7bdat for input
__________Using the **KEEP** statement__________

SAS Codes (E45a)

```
LIBNAME test1 'c:\sas_exercise';
DATA test1.newset1;
SET test1.olddata (KEEP= State population drivers);
RUN;
PROC PRINT DATA=test1.newset1;
RUN;
```

LIBNAME test1 "c:\sas_exercise" creates a library named test1 representing the folder of c:\sas_exercise.

DATA test1.newset1 creates a permanent SAS dataset called newset1.sas7bdat and stores it in the test1 library.

SET test1.olddata (KEEP=State population drivers) instructs SAS to use the dataset olddata.sas7bdat stored in the test1 library to create the newset1.sas7bdat dataset by only retaining variables of State, population, and drivers.

RUN kicks off the DATA step.

PROC PRINT DATA=test1.newset1 tells SAS to print the data named newset1.sas7bdat stored in the test1 library.

RUN kicks off the PROC PRINT step.

Using the **DROP** statement

SAS Codes (E45b)

```
LIBNAME test1 'c:\sas_exercise';
DATA test1.newset1;
SET test1.olddata (drop= region hwyfuel);
RUN;
PROC PRINT DATA=test1.newset1;
RUN;
```

LIBNAME test1 "c:\sas_exercise" creates a library named test1 representing the folder of c:\sas_exercise.

DATA test1.newset1 creates a permanent SAS dataset called newset1.sas7bdat and stores it in the test1 library.

SET test1.olddata (drop=region hwyfuel) instructs SAS to use the dataset olddata.sas7bdat stored in the test1 library to create the newset1.sas7bdat dataset. Besides, the drop option instructs SAS to drop the variables region and hwyfuel from the newset1.sas7bdat file

RUN kicks off the DATA step.

PROC PRINT DATA=test1.newset1 tells SAS to print the data named newset1.sas7bdat stored in the test1 library.

RUN kicks off the PROC PRINT step.

Using the **RENAME** statement

SAS Codes (E45c)

```
LIBNAME test1 'c:\sas_exercise';
DATA test1.newset1 (RENAME = (Region=LargeZone
Drivers=LicensedDrivers));
SET test1.olddata (drop= hwyfuel);
RUN;
PROC PRINT DATA=test1.newset1;
RUN;
```

LIBNAME test1 "c:\sas_exercise" creates a library named test1 representing the folder of c:\sas_exercise.

DATA test1.newset1 creates a permanent SAS dataset called newset1 and stores it in the test1 library. In addition, the (RENAME=(Region=LargeZone Drivers=LicensedDrivers) option tells SAS to rename the Region variable with the new name of LargeZone, and Drivers with the new name of LicensedDrivers in the newset1 dataset to be created.

SET test1.olddata (drop=hwyfuel) instructs SAS to use the dataset olddata.sas7bdat stored in the test1 library to create the newset1.sas7bdat dataset. Besides, the drop option instructs SAS to drop the variable hwyfuel from the newset1.sas7bdat file.

RUN kicks off the DATA step.

PROC PRINT DATA=test1.newset1 tells SAS to print the data named newset1.sas7bdat stored in the test1 library.

The example below is using the .csv data for input

SAS Codes (E45d)

```
LIBNAME test1 'c:\sas_exercise';
DATA test1.newset (RENAME = (vehicles=RegisteredVehicles
Drivers=LicensedDrivers));
   DROP region HWYfuel;
INFILE  'c:\sas_exercise\olddata.csv' DLM=','  FIRSTOBS=2  DSD
MISSOVER;
INPUT region :$13. State $ population vehicles drivers HWYfuel;
RUN;

PROC PRINT DATA=test1.newset;
RUN;
```

LIBNAME test1 "c:\sas_exercise" creates a library named test1 representing the folder of c:\sas_exercise.

DATA test1.newset creates a permanent SAS dataset called newset and stores it in the test1 library. Also, the (RENAME=(vehicles=RegisteredVehicles Drivers=LicensedDrivers) option tells SAS to rename the vehicles variable with a new name of RegisteredVehicles, and Drivers with a new name of LicensedDrivers in the newset dataset to be created.

The DROP region HWYfuel tells SAS to drop the variables region and HwyFuel from the new newset.sas7bdat file.

INFILE "c:\sas_exercise\olddata.csv" specifies the directory path and the file olddata.csv to be read and used to create the newset.sas7bdat file.

DLM=',' tells SAS that the delimiter used in the olddata.csv file is the , comma.

FIRSTOBS=2 tells SAS that the SAS should start to read data from row 2.

DSD further instructs SAS to handle three things:

1: Treat two consecutive delimiters (, ,) as a missing value.

2: Remove quotation marks from character values and treat everything within a pair of quotation marks as a single observation for the variable (e.g., "AB, C", XYZ will be treated as 2 variables, not 3. The "AB, C" is treated as 1).

3: Assume observations for variables are separated by the , comma.

MISSOVER prevents SAS from going to the next data line if it can't locate values with the current dateline for all variables listed in the INPUT statement. SAS assigns a missing value for variables without values.

INPUT region :$13. State $ population vehicles drivers HWYfuel instructs SAS to (a) read up to 13 columns for the character variable region (Colon :$n means up to the nth columns or until a separator is encountered) and (b) treat State as a character variable, and population, vehicles, drivers, and HWYfuel as numerical variables.

RUN kicks off the DATA step.

PROC PRINT DATA=test1.newset tells SAS to print the data named newset.sas7bdat stored in the test1 library.

RUN kicks off the PROC PRINT step.

4.6 CREATING NEW DATA VARIABLES

It is almost certain that you will need to create new variables when analyzing your data. Creating new data variables occurs in the DATA step. There are many ways to accomplish the task, including but not limited to simple assignment, SAS functions, subsetting (IF-THEN statements), concatenation, substrings, and the ANY functions.

4.6.1 Simple Assignment

The simple assignment method, as illustrated below, is one of the most common ways to create new variables.

Example: 4.6.1

Data: boat.csv as illustrated below

State	Boats	LandArea	WaterArea
AK	1234569	570,640.95	94,743.10
TX	1578989	261,231.71	7,364.75
CA	2156983	155,779.22	7,915.52
MT	15693	145,545.80	1,493.91
NM	19879	121,298.15	292.15
AZ	13568	113,594.08	396.22
NV	6356	109,781.18	790.65
CO	65893	103,641.89	451.78
OR	45612	95,988.01	2,390.53
WY	6325	97,093.14	719.87

Task: Create new variables listed below:

1. TotalArea (LandArea+waterArea)
2. Land Area/WaterArea ratio (LandArea/WaterArea)
3. Boats Per Square Miles (Boats/TotalArea)
4. BoatOwnership Rate – low, medium, or high (# of boat/square miles< 1 is low; 1<= # of boat/square miles<20 is medium; # of boat/square miles=>20 is high)
5. Status based on Land/Water ratio – Dry or Water (when the land/water ratio>20, it is "Dry"; otherwise, it is "Water."

SAS Codes (E461)

```
DATA newBoat  ;
FORMAT BoatOwnership $7. Status $5.;
INFORMAT LandArea comma10.2 WaterArea comma9.2;
INFILE  'c:\sas_exercise\boat.csv' DLM=','  FIRSTOBS=2  DSD
MISSOVER;
INPUT State $ boats LandArea WaterArea;

TotalArea = LandArea + waterArea;
LandWaterR = LandArea/WaterArea;
BoatPerSM = boats/waterArea;

IF BoatPerSM <1 then BoatOwnership = 'Low';
ELSE IF (boatPerSM =>1 and boatPerSM< 20.0) THEN BoatOwnership=
'Medium';
ELSE IF boatPerSM=>20.0 THEN BoatOwnership= 'High';
ELSE IF boatperSM =. THEN BOATOwnership= 'Unknown';
IF LandWaterR>20 then Status='Dry';
IF LandWaterR=<20 then Status='Water';
RUN;

*Below is the 1st Proc Print;
PROC PRINT DATA=newboat (obs=11);
RUN;
DATA newboat2;
RETAIN State Boats LandArea WaterArea TotalArea LandWaterR
BoatPerSM BoatOwnership Status;
SET newboat;
RUN;
*Below is the 2nd Proc Print;
PROC PRINT DATA=newboat2 (Obs=10);
RUN;
```

DATA newBoat creates a new SAS dataset called newBoat in the default WORK library.

FORMAT BoatOwnership $7. Status $5. defines BoatOwnership as a 7-column character width variable, and Status is a 5-column width character variable (keep in mind FORMAT only tells SAS how to display a variable and will not affect the actual data).

INFORMAT LandArea comma10.2 WaterArea comma9.2; tells SAS that (a) LandArea has the thousands separator (comma) with 2 decimal points for a total width of 10 columns, and (b) WaterArea has the thousands separator (comma) and it has 2 decimal points for a total width of 9 columns.

INFILE "c:\sas_exercise\boat.csv" specifies the directory path for the file boat.csv to be read from and used to create the newBoat.sas7bdat file.

DLM=',' tells SAS that the delimiter used in the boat.csv file is , comma.

FIRSTOBS=2 tells SAS to read data from row 2.

DSD further instructs SAS to handle three things:

1: Treat two consecutive delimiters (, ,) as a missing value.

2: Remove quotation marks from character values and treat everything within a quotation mark together as a single observation for the variable (e.g., "AB, C", XYZ will be treated as 2 variables, not 3. The "AB, C" is treated as 1).

3: Assume data separator is the , comma.

MISSOVER prevents SAS from going to the next data line if it can't locate values with the current dateline for all variables listed in the INPUT statement. SAS assigns a missing value for variables without values.

INPUT State $ boats LandArea WaterArea instructs SAS to (a) read the character variable State and the numerical variables of boats, LandArea, and WaterArea in the boat.csv file.

TotalArea=LandArea+waterArea is an assignment statement which creates a new variable TotalArea for the new dataset newBoat.sas7bdat.

LandWaterR=LandArea/WaterArea is another assignment statement creating a new variable of LandWaterR.

BoatPerSM=boats/waterArea is an assignment statement creating a new variable called BoatPerSM by existing variables.

IF BoatPerSM<1 then BoatOwnership="Low" is a conditional statement used to create a new variable called BoatOwnership and fill the observations with Low if the specified condition BoatPerSM<1 is met.

ELSE IF … THEN continues the boatPerSM conditional scenarios by assigning different values to the BoatOwnership variable.

IF LandWaterR>20 then Status="Dry" is a conditional statement used to create a new variable called Status and fills observations with Dry when the specified condition is met.

IF LandWaterR=<20 then Status="Water" is a conditional statement used to create a new variable called Status and fills observations with Water when the specified condition is met.

Keep in mind that all these newly created variables are part of the newBoat.sas7bdat file.

RUN kicks off the DATA step.

PROC PRINT DATA=newboat (obs=11) instructs SAS to print the first 11 rows in the newly created file called newBoat.sas7bdat stored in the default WORK library.

DATA newboat2 creates a new SAS dataset named newboat2 residing in the default WORK library.

The RETAIN … statement lists all variables in a different order with their basic INFORMATs (e.g., $) specified. It rearranges the order of variables.

SET newboat tells SAS that the newboat2 file is to be based on the newboat.sas7bdat file in the default WORK library.

RUN kicks off the DATA step.

PROC PRINT DATA =newboat2 (Obs=10) directs SAS to print the first 10 rows of the newboat2.sas7bdat file in the WORK library.

RUN kicks off the PROC PRINT step.

See output variable order differences per the FORMAT statement used.

Result from the 1st PROC PRINT DATA=newboat

Default variable order

Obs	BoatOwnership	Status	landarea	waterarea	State	boats	TotalArea	LandWaterR	BoatPerSM
1	Medium	Water	570640.95	94743.10	AK	1234569	665384.05	6.023	13.031
2	High	Dry	261231.71	7364.75	TX	1578989	268596.46	35.471	214.398
3	High	Water	155779.22	7915.52	CA	2156983	163694.74	19.680	272.500
4	Medium	Dry	145545.80	1493.91	MT	15693	147039.71	97.426	10.505
5	High	Dry	121298.15	292.15	NM	19879	121590.30	415.191	68.044
6	High	Dry	113594.08	396.22	AZ	13568	113990.30	286.694	34.244
7	Medium	Dry	109781.18	790.65	NV	6356	110571.83	138.849	8.039
8	High	Dry	103641.89	451.78	CO	65893	104093.67	229.408	145.852
9	Medium	Dry	95988.01	2390.53	OR	45612	98378.54	40.153	19.080
10	Medium	Dry	97093.14	719.87	WY	6325	97813.01	134.876	8.786
11	Low	Water	56538.90	40174.61	WI	25694	96713.51	1.407	0.640

Result from the 2nd PROC PRINT Data=newboat2

Re-arranged order of variables

The SAS System

Obs	State	Boats	LandArea	WaterArea	TotalArea	LandWaterR	BoatPerSM	BoatOwnership	Status
1	AK	1234569	570640.95	94743.10	665384.05	6.023	13.031	Medium	Water
2	TX	1578989	261231.71	7364.75	268596.46	35.471	214.398	High	Dry
3	CA	2156983	155779.22	7915.52	163694.74	19.680	272.500	High	Water
4	MT	15693	145545.80	1493.91	147039.71	97.426	10.505	Medium	Dry
5	NM	19879	121298.15	292.15	121590.30	415.191	68.044	High	Dry
6	AZ	13568	113594.08	396.22	113990.30	286.694	34.244	High	Dry
7	NV	6356	109781.18	790.65	110571.83	138.849	8.039	Medium	Dry
8	CO	65893	103641.89	451.78	104093.67	229.408	145.852	High	Dry
9	OR	45612	95988.01	2390.53	98378.54	40.153	19.080	Medium	Dry
10	WY	6325	97093.14	719.87	97813.01	134.876	8.786	Medium	Dry

4.6.2 SAS Functions

SAS itself has hundreds of preprogrammed functions that you can use to convert and compute data. For example, the **LOG (variable)** function generates the natural logarithmic value of an observation. The **WEEKDAY (var)** function deciphers which day of the week an observation is. The **UPCASE (variable)** function converts all character observations into uppercase. The examples below illustrate how SAS functions are used to create new variables.

Example: 4.6.2

Data: speed.csv as illustrated below

Location	RoadName	Direction	RoadType	MeasureDate	AMspeed	PMspeed
PG569	I95	N	1	1/1/19	35	45
PG569	I95	N	1	1/2/19	42	71
PG569	I95	N	1	1/3/19	32	70
PG569	I95	N	1	1/4/19	58	55
PG569	I95	N	1	1/5/19	72	78
PG569	I95	N	1	1/6/19	74	72
PG569	I95	N	1	1/7/19	60	64
PG569	I95	N	1	1/8/19	36	62
PG569	I95	N	1	1/9/19	45	62
PG569	I95	N	1	1/10/19	28	32

Task: Create new variables as listed below:

1. Day of Week (DOW) (Monday, Tuesday… Sunday) based on MeasureDate
2. A variable (WKWD) to differentiate weekend from weekday based on MeasureDate
3. Maximum speed (MS) between the AM and PM speeds
4. The average speed (AvS) between the AM and PM speeds.

SAS Codes (E462)

```
DATA myspeed2 (RENAME = (MeasureDate=MD RoadName=RN Direction=D
RoadType=RT));
INFORMAT location $5. MeasureDate mmddyy10. ;
INFILE  'c:\sas_exercise\speed_chapter4.csv' DLM=',' FIRSTOBS=2
DSD MISSOVER;
INPUT Location $ RoadName $ Direction $ RoadType  MeasureDate
AMspeed PMspeed;

DOW= WEEKDAY(MeasureDate);
IF (DOW= 1 or DOW=7) THEN WKWD='weekend';
ELSE WKWD='Weekday';

MS= MAX(AMspeed,PMspeed);
```

```
AvS= MEAN(AMspeed,PMspeed);

RUN;
PROC PRINT DATA=myspeed2 (obs=19);
RUN;
```

DATA myspeed2 creates a new SAS dataset named myspeed2.sas7bdat and stores it in the default WORK library. Variable MeasureDate is renamed as MD, RoadName as RN, Direction to D, and RoadType to RT.

INFORMAT location $5. MeasureDate mmddyy10. defines that location is a character variable with a width of 5 columns, and MeasureDate is a date variable with the INFORMAT of mmddyy10.

INFILE "c:\sas_exercise\speed_chapter4.csv" specifies the directory path and the file speed_chapter4.csv is to be read and used to create the myspeed2.sas7bdat file.

DLM=',' tells SAS that the delimiter used in the speed_chapter4.csv file is the , comma.

FIRSTOBS=2 tells SAS that the SAS should start to read data from row 2.

DSD further instructs SAS to handle three things:

1: Treat two consecutive delimiters (, ,) as a missing value.

2: Remove quotation marks from character values and treat everything within a quotation mark together as a single observation for the variable (e.g., "AB, C", XYZ will be treated as 2 variables, not 3. The "AB, C" is treated as 1).

3: Assume data separator is the , comma.

MISSOVER prevents SAS from going to the next data line if it can't locate values with the current dateline for all variables listed in the INPUT statement.

INPUT ... statement lists all variables and their INFOMRATs (e.g., $) corresponding to the speed_chapter4.csv file.

DOW=WEEKDAY(MeasureDate) is an assignment statement creating a new variable named DOW by using SAS function Weekday (MeasureDate). The SAS Weekday function determines the day of week for an observation (Sunday is Day 1, Monday is Day 2... Saturday is Day 7).

IF (DOW=1 or DOW=7) Then WKWD="weekend" is a conditional statement used to create a new variable named WKWD. When the specified condition is met, WKWD is assigned weekend.

ELSE WKWD is assigned Weekday.

MS=MAX(AMspeed,PMspeed) creates a new variable called MS by using the MAX function. The MAX(AMspeed,PMspeed) assigns the bigger value between AMspeed and PMspeed to the MS variable.

AvS=MEAN(AMspeed,PMspeed) creates a new variable called AvS by using the SAS MEAN function. The MEAN(AMspeed,PMspeed) computes the average of the AMspeed and PMspeed and assigns the average to the AvS variable.

RUN kicks off the DATA step.

PROC PRINT DATA=myspeed2 (obs=19) instructs SAS to print the first 11 rows of the myspeed2.sas7bdat in the WORK library on the screen.

RUN kicks off the PROC PRINT step.

Result:
The result from the PROC PRINT Data=myspeed2

Obs	location	MD	RN	D	RT	AMspeed	PMspeed	DOW	WKWD	MS	AvS
1	PG569	21550	I95	N	1	35	45	3	Weekday	45	40.0
2	PG569	21551	I95	N	1	42	71	4	Weekday	71	56.5
3	PG569	21552	I95	N	1	32	70	5	Weekday	70	51.0
4	PG569	21553	I95	N	1	58	55	6	Weekday	58	56.5
5	PG569	21554	I95	N	1	72	78	7	weekend	78	75.0
6	PG569	21555	I95	N	1	74	72	1	weekend	74	73.0
7	PG569	21556	I95	N	1	60	64	2	Weekday	64	62.0
8	PG569	21557	I95	N	1	36	62	3	Weekday	62	49.0
9	PG569	21558	I95	N	1	45	62	4	Weekday	62	53.5
10	PG569	21559	I95	N	1	28	32	5	Weekday	32	30.0
11	PG569	21560	I95	N	1	36	46	6	Weekday	46	41.0
12	PG569	21561	I95	N	1	70	68	7	weekend	70	69.0
13	PG569	21562	I95	N	1	71	73	1	weekend	73	72.0
14	PG569	21563	I95	N	1	56	60	2	Weekday	60	58.0
15	PG569	21564	I95	N	1	46	60	3	Weekday	60	53.0
16	PG569	21565	I95	N	1	38	42	4	Weekday	42	40.0
17	PG569	21566	I95	N	1	36	24	5	Weekday	36	30.0
18	PG569	21567	I95	N	1	24	45	6	Weekday	45	34.5
19	PG569	21569	I95	N	1	66	78	1	weekend	78	72.0

4.6.3 If-Then, If-Then-Else, and Where

You can use the IF-THEN and IF-THEN-ELSE logic statements during your DATA step to create new variables. You can also use the WHERE logic function during your DATA step (WHERE can also be used with the PROC procedure), but be aware of the two special cases with which the WHERE statement doesn't work:

- Your subsetting condition uses new variables created within your DATA step.
- WHERE is used with your INPUT statement.

Example: 4.6.3

Data: speed_chapter4.csv

Task: Create a new variable called Congestion based on the speed profile (green=>45 mph, 30 mph=<yellow<45mph, red<30 mph.

SAS Codes (E463)

```
DATA myspeed  ;
INFORMAT location $5. MeasureDate mmddyy10. ;
INFILE  'c:\sas_exercise\speed_chapter4.csv' DLM=',' FIRSTOBS=2
DSD MISSOVER;
INPUT Location $ RoadName $ Direction $ RoadType  MeasureDate
AMspeed PMspeed;
RUN;
PROC PRINT data=myspeed;
RUN;
FORMAT Congestion $8.;

DATA myspeed2 ;
SET myspeed;
IF AMspeed=>45 THEN Congestion = 'Green';
ELSE IF 30<=AMspeed <45 THEN Congestion= 'Yellow';
ELSE IF AMspeed <30 THEN Congestion='Red';
RUN;

PROC PRINT DATA=myspeed2 (Obs=11);
RUN;
```

DATA myspeed creates a new SAS dataset named myspeed.sas7bdat and stores it in the default WORK library.

INFORMAT location $5. MeasureDate mmddyy10. defines that location is a character variable with a width of 5 columns, and MeasureDate is a date variable with the INFORMAT of mmddyy10.

INFILE "c:\sas_exercise\speed_chapter4.csv" specifies the directory path and the file speed_chapter4.csv to be read and used to create the myspeed.sas7bdat file.

DLM=',' tells SAS that the delimiter used in the speed_chapter4.csv file is , comma.

FIRSTOBS=2 tells SAS that the SAS should start to read data from row 2.

DSD further instructs SAS to handle three things:

1: Treat two consecutive delimiters (, ,) as a missing value.

2: Remove quotation marks from character values and treat everything within a quotation mark together as a single observation for the variable (e.g., "AB, C ", XYZ will be treated as 2 variables, not 3. The "AB, C" is treated as 1).

3: Assume data separator is the , comma.

MISSOVER prevents SAS from going to the next data line if it can't locate values with the current dateline for all variables listed in the INPUT statement.

INPUT … statement lists all variables and their INFOMRAT (e.g., $) corresponding to the speed_chapter4.csv file.

RUN kicks off the DATA step.

PROC PRINT data=myspeed instructs SAS to print the myspeed.sas7bdat in the WORK library on the screen.

DATA myspeed2 creates a new SAS dataset named myspeed2.sas7bdat and stores it in the default WORK library.

SET myspeed tells SAS that the myspeed.sas7bdat in the WORK library is used to create the myspeed2 data file.

IF AMspeed=>45 THEN Congestion="Green" is a conditional statement used to create a new variable called Congestion. When AMspeed=45 or >45, the Congestion gets assigned a value of Green.

ELSE IF … THEN continues other conditions and assigns corresponding values to the Congestion variable.

RUN kicks off the DATA step.

PROC PRINT DATA=myspeed2 (Obs=11) instructs SAS to print the first 11 rows of the myspeed2.sas7bdat dataset on the screen.

RUN kicks off the PROC PRINT step.

Result:

Obs	location	MeasureDate	RoadName	Direction	RoadType	AMspeed	PMspeed	Congestion
1	PG569	21550	I95	N	1	35	45	Yellow
2	PG569	21551	I95	N	1	42	71	Yellow
3	PG569	21552	I95	N	1	32	70	Yellow
4	PG569	21553	I95	N	1	58	55	Green
5	PG569	21554	I95	N	1	72	78	Green
6	PG569	21555	I95	N	1	74	72	Green
7	PG569	21556	I95	N	1	60	64	Green
8	PG569	21557	I95	N	1	36	62	Yellow
9	PG569	21558	I95	N	1	45	62	Green
10	PG569	21559	I95	N	1	28	32	Red
11	PG569	21560	I95	N	1	36	46	Yellow

4.6.4 Concatenation – CAT, CATS, CATT, and CATX

You can use concatenation statements (CAT, CATS, CATT, and CATX) to create new variables by joining existing string variables. Below are four different concatenation types.

- newvar=**CAT**(var1, var2…)
  Joining variables (observations) together without modifying original observations.
- newvar=**CATS**(var1, var2…)
  Joining variables (observations) after stripping all trailing and leading blanks from the original variables.
- newvar=**CATT**(var1,var2…)
  Joining variables (observations) after stripping only trailing blanks from the original variables.

- newvar=**CATX**(",", var1,var2)
  Joining observations by first stripping both leading and trailing blanks and then adding one or more characters between them (in the example here, a comma , is added between variables).

Example: 4.6.4

Data:
my_concatenation.sas7bdat

Notice the heading and trailing blanks for observations with the variable of LocationID illustrated below.

StateName	LocationID	Amspeed	Pmspeed
NC	A3067	49	44
NC	B9187	48	38
NC	B6790	46	36
VA	A11990	46	42
VA	A11785	47	42
VA	A34569	48	48
VA	A98127	41	41
MD	M9878	42	38
MD	N8972	41	42
DE	A5678	45	36
PA	B9712	44	45
PA	B5637	47	44
PA	B3721	42	38
PA	C6789	41	37
NJ	B8965	38	40
NJ	A6712	42	43

Task: Create 4 different new IDs by concatenating the "Statename" and "LocationID" through **CAT, CATS, CATT,** and **CATX** statements.

SAS Code (E464)

```
LIBNAME mycat 'c:\sas_exercise';
DATA catData;
SET mycat.cat;
ID1_cat=CAT(Statename, LocationID);
ID2_cats=CATS(Statename,LocationID);
ID3_catt=CATT(Statename,LocationID);
ID4_catx=CATX('_', Statename,LocationID);

RUN;

PROC PRINT DATA = catdata;
RUN;
```

LIBNAME mycat "c:\sas_exercise" defines the mycat library representing the folder of c:\sas_exercise.

DATA catData creates a new dataset called catData and stores it in the default WORK library.

SET mycat.cat instructs SAS to use the cat.sas7bdat in the mycat library to create the new catData.

ID1_cat=CAT(statename, LocationID) creates a new variable called ID1_cat by using the SAS command CAT to join variables Statename and LocationID without any modification to them.

ID2_cats=CATS(Statename,LocationID) creates a new variable called ID2_cats by using the SAS command CATS to join variables Statename and LocationID after stripping all trailing and leading blanks of variables Statename and LocationID.

ID3_catt=CATT(Statename,LocationID) creates a new variable called ID3_catt by using the SAS command CATT to join variables Statename and LocationID after stripping trailing blanks of variables Statename and LocationID.

ID4_catx=CATX('_', Statename,LocationID) creates a new variable called ID4_catx by using the SAS command CATX to join variables Statename and LocationID after stripping trailing blanks of variables Statename and LocationID. Besides, the symbol _ as specified by "_" is inserted between the variables Statename and LocationID.

RUN kicks off the DATA step.

PROC PRINT DATA=catData instructs SAS to print the catdata.sas7bdat dataset on the screen.

RUN kicks off the PROC PRINT step.

Result:

StateName	LocationID	Am
NC	A3067	
NC	B9187	
NC	B6790	
VA	A11990	
VA	A11785	
VA	A34569	
VA	A98127	
MD	M9878	
MD	N8972	
DE	A5678	
PA	B9712	
PA	B5637	
PA	B3721	
PA	C6789	
NJ	B8965	
NJ	A6712	

Obs	StateName	LocationID	Amspeed	Pmspeed	ID1_cat	ID2_cats	ID3_catt	ID4_catx
1	NC	A3067	49	44	NCA3067	NCA3067	NCA3067	NC_A3067
2	NC	B9187	48	38	NC B9187	NCB9187	NC B9187	NC_B9187
3	NC	B6790	46	36	NCB6790	NCB6790	NCB6790	NC_B6790
4	VA	A11990	46	42	VA A11990	VAA11990	VA A11990	VA_A11990
5	VA	A11785	47	42	VA A11785	VAA11785	VA A11785	VA_A11785
6	VA	A34569	48	48	VA A34569	VAA34569	VA A34569	VA_A34569
7	VA	A98127	41	41	VA A98127	VAA98127	VA A98127	VA_A98127
8	MD	M9878	42	38	MDM9878	MDM9878	MDM9878	MD_M9878
9	MD	N8972	41	42	MDN8972	MDN8972	MDN8972	MD_N8972
10	DE	A5678	45	36	DE A5678	DEA5678	DE A5678	DE_A5678
11	PA	B9712	44	45	PAB9712	PAB9712	PAB9712	PA_B9712
12	PA	B5637	47	44	PA B5637	PAB5637	PA B5637	PA_B5637
13	PA	B3721	42	38	PAB3721	PAB3721	PAB3721	PA_B3721
14	PA	C6789	41	37	PA C6789	PAC6789	PA C6789	PA_C6789
15	NJ	B8965	38	40	NJB8965	NJB8965	NJB8965	NJ_B8965
16	NJ	A6712	42	43	NJ A6712	NJA6712	NJ A6712	NJ_A6712

4.6.5 Substring (Substr) and Length

The **SUBSTR** statement is a very powerful function for you to use in a wide range of ways, including the creation of new variables within existing variables. The **SUBSTR** statement can be used alone or with statements to extract a portion of a string variable. It takes the form of:

new_var=**SUBSTR**(original_var, start, digit)

where

- new_var: the name of the new variable you are creating.
- original_var: the variable name from which you are extracting a portion.
- Start: column # where extraction starts. It must be a numeric value (e.g., 4).
- Digit: # of columns to be extracted from the Start position (e.g., the start=3, digit =5 specification leads to extracting columns from 3 to 7 for a total of 5 columns). Digit is optional. Without it, the extraction will retain all columns to the right from the start position.

The **SUBSTR** statement can also be used on the "left" side of a statement, without being assigned to a new variable. When it is used on the left side, it modifies the original observations.

Example: 4.6.5a

SAS Codes (E465a)

```
Data new;
var1= "abcdefghi";
var2= "abcdefghi";
var3= "abcdefghi";
newvar1=substr(var1,2,4);
newvar2=substr(var2,5);
```

*the 3 statements below use the Substr on the left;

```
SUBSTR(var1, 2,1)= "B";
SUBSTR(var2, 2,3) ="x13";
SUBSTR(var3, 2,6)="BB";
RUN;
PROC PRINT data=new;
RUN;
```

Data new creates a new dataset named new in the default WORK library.

var1="abcdefghi"
var2="abcdefghi"
var3="abcdefghi"
assign 3 new variables called var1, var2, and var3 for the new dataset with the value of abcdefghi.

newvar1=substr(var1,2,4) instructs SAS to create a new variable called newvar1 by extracting (substr) a portion of the var1 variable. The extraction starts at column 2 and has a total width of 4 columns, meaning starting at the 2nd column and stopping after the 5th column is extracted.

newvar2=(substr(var2,5) instructs SAS to create a new variable called newvar2 by extracting (substr) a portion of the var2 variable. The extraction starts at column 5 and retains all other columns to the right of the 5th columns which var2 has.

SUBSTR(var1,2,1)="B" statement instructs SAS to modify the original variable var1 by replacing a portion of the var1 with B. The replacement starts with column 2 and only 1 column width.

SUBSTR(var2, 2,3)="x13" statement instructs SAS to modify the original variable var2 by replacing a portion of the var2 with x13. The replacement starts with column 2 for a total of 3 columns.

SUBSTR(var3, 2,6)="BB" statement instructs SAS to modify the original variable var3 by replacing a portion of the var3 with BB. The replacement starts with column 2 for a total of 6 columns.

RUN kicks off the Data step.

PROC PRINT data=new instructs SAS to print the first 11 rows of the new.sas7bdat dataset on the screen.

RUN kicks off the PROC PRINT step.

Result:

Obs	var1	var2	var3	newvar1	newvar2
1	aBcdefghi	ax13efghi	aBB hi	bcde	efghi

Illustration:

	Column Number								
	1	2	3	4	5	6	7	8	9
var1	a	b	c	d	e	f	g	h	I
var2	a	b	c	d	e	f	g	h	I
var3	a	b	c	d	e	f	g	h	I
SUBSTR(var1, 2,1)= "B";	a	B	c	d	e	f	g	h	I
SUBSTR(var2, 2,3) ="x13";	a	x	1	3	e	f	g	h	I
SUBSTR(var3, 2,6)="BB";	a	B	B					h	I

	Column Number								
	1	2	3	4	5	6	7	8	9
var1	a	b	c	d	e	f	g	h	I
newvar1=substr(var1,2,4)		b	c	d	e				
newvar2=substr(var1, 5)					e	f	g	h	I

Using the SUBSTR with the LENGTH Statement

The LENGTH statement tells SAS the number of columns (width) an observation occupies. It takes the generic form as listed below.

widthOFvar1=LENGHT(var1);

The above statement uses the **LENGTH** statement to count how many columns each observation occupies by the variable var1.

Example: 4.6.5b

SAS Codes (E465b)

```
Data new;
var1= "a";
var2= "ab";
var3= "abcdef";
L_var1=length(var1);
L_var2=length(var2);
L_var3=length(var3);
RUN;
PROC PRINT data=new;
RUN;
```

Data new creates a new dataset named new in the default WORK library.
var1="a"
var2="ab"
var3="abcdef"
assigns three new variables called var1, var2, and var3 for the new dataset with the values of a, ab, and abcdef, respectively.
L_var1=length(var1) creates a new variable called L_var1 by using the SAS function length The length(var1) counts numbers of columns var1 observations have.
L_var2=length(var2) creates a new variable called L_var2 by using the SAS function lenght. The length(var2) counts numbers of columns var2 observations have.
L_var3=length(var3) creates a new variable called L_var3 by using the SAS function length. The length(var3) counts numbers of columns var3 observations have.
RUN kicks off the Data step.
PROC PRINT data=new instructs SAS to print the first 11 rows of the new.sas7bdat dataset on the screen.
RUN kicks off the PROC PRINT step.

Result:

Obs	var1	var2	var3	L_var1	L_var2	L_var3
1	a	ab	abcdef	1	2	6

Example: 4.6.5c

Data: substring_exer.sas7bdat

COuntyID	RoadName	Amspeed	Pmspeed
AlexanderNC	US26	49	44
AnsonNC	US17	48	38
CamdenNC	SR180	46	36
CarolineVA	US17	46	42
ChesterfieldVA	SR200	47	42
CulpepperVA	SR409	48	48
FauquierVA	CR18	41	41
MontgomeryMD	CR192	42	38
BaltimoreMD	CR192	41	42
New CastleDE	US17	45	36
BerksPA	US202	44	45
CarbonPA	SR120	47	44
UnionNJ	CR135	42	43
ColumbiaPA	CR480	42	38
LancasterPA	CR201	41	37
AtlanticNJ	CR19	38	40

Task:

1. Extract the first 2 characters and the numerical part of the RoadName to create two new variables called Class and Num.
2. Create a new variable called State based on the last 2 columns of the CountyID.
3. Replace the 2nd to 4th columns of the variable CountyID with "_ &_."

SAS Codes (E465c)

```
LIBNAME myroad 'c:\sas_exercise';
DATA new1 (DROP=Amspeed Pmspeed);
SET myroad.substring_exer ;
Class=SUBSTR(RoadName,1,2);
Num1=INPUT(SUBSTR(RoadName, 3, 3), 3.);
Num2=INPUT(SUBSTR(RoadName,3), 3.);
State=SUBSTR(countyID,LENGTH(countyID)-1,2);
CountyID2=CountyID;
SUBSTR(CountyID2, 2, 4)= '_&_';
RUN;
PROC PRINT DATA=new1;
RUN;
```

LIBNAME myroad "c:\sas_exercise" defines a library myroad representing the folder c:\sas_exercise.

DATA new1 creates a new SAS dataset called new1 in the default WORK library. The DROP command requests the variables Amspeed and Pmspeed be dropped from the new dataset.

SET myroad.substring_exer instructs SAS to use the file substring_exer.sas7bdat in the myroad library to create the new1 dataset.

Class=SUBSTR(RoadName,1,2) creates a new variable named Class by using the SUBSTR function to extract from the variable RoadName starting at column 1 (with 2 columns width) and stopping at 2.

Num1 =INPUT(SUBSTR(RoadName, 3,3),3.) creates a new variable named Num1 and it is a three-digit integer numerical variable assigned by the INPUT (, 3.) through extracting (SUBSTR(RoadName, 3,3) of the variable RoadName from the 3rd column and stops after the 5th column (3 columns wide) is extracted.

Num2 =INPUT(SUBSTR(RoadName, 3),3.) creates a new variable named Num2 and it is a three-digit integer numerical variable assigned by the INPUT (, 3.) through extracting (SUBSTE(RoadName, 3) of the variable RoadName from the 3rd column and all digits after.

State=SUBSTR(countyID,LENGTH(countyID)-1,2) creates a new variable named State by using the SAS function SUBSTR to extract countyID starting at the column number of (column width minus 1) with column width being determined by another SAS function LENGTH(countyID). The variable State is 2 column wide specified by the State=SUBSTR(…,2)

For example:

Length(countyID)=Length(Leon)=4

Length(countyID) -1=4 – 1=3

SUBSTR(countyID,LENGTH(countyID)-1,2)=SUBSTR(Leon,3,2)=bs

CountyID2=CountyID creates a new variable CountyID2 by simply assigning CountyID to it.

SUBSTR(CountyID2, 2,4)='_&_' modifies the CountyID2 variable by replacing the 2nd column to the 5th (4 columns wide starting with the 2nd column and ending after the 5th column) column by the symbol of _&_.

RUN kicks off the DATA step.

PROC PRINT Data=new1 tells SAS to print the new1.sas7bdat in the WORK library on the screen.

RUN kicks off the PROC PRINT procedure.

Result:

Obs	COuntyID	RoadName	F5	F6	Class	Num1	Num2	State	CountyID2
1	AlexanderNC	US26			US	26	26	NC	A_&_nderNC
2	AnsonNC	US17			US	17	17	NC	A_&_NC
3	CamdenNC	SR180			SR	180	180	NC	C_&_nNC
4	CarolineVA	US17			US	17	17	VA	C_&_ineVA
5	ChesterfieldVA	SR200			SR	200	200	VA	C_&_erfieldVA
6	CulpepperVA	SR409			SR	409	409	VA	C_&_pperVA
7	FauquierVA	CR18			CR	18	18	VA	F_&_ierVA
8	MontgomeryMD	CR192			CR	192	192	MD	M_&_omeryMD
9	BaltimoreMD	CR192			CR	192	192	MD	B_&_moreMD
10	New CastleDE	US17			US	17	17	DE	N_&_astleDE
11	BerksPA	US202			US	202	202	PA	B_&_PA
12	CarbonPA	SR120			SR	120	120	PA	C_&_nPA
13	UnionNJ	CR135			CR	135	135	NJ	U_&_NJ
14	ColumbiaPA	CR480			CR	480	480	PA	C_&_biaPA
15	LancasterPA	CR201			CR	201	201	PA	L_&_sterPA
16	AtlanticNJ	CR19			CR	19	19	NJ	A_&_ticNJ

4.6.6 ANY Functions – Locating the Starting Point of a Substring

The ANY function includes **ANYALNUM, ANYALPHA, ANYDIGIT**, and **ANYSPACE**. These commands determine the location of the first occurrence of a letter/number, a letter, a number, or space in an observation, and then return that location number, respectively.

You can use the location of the first occurrence to guide the substring action in creating new variables.

Example: 4.6.6

Data:
anyspace2.sas7bdat

DATETIME
1/1/2017 1:00
1/1/2017 20:00
1/12/2017 23:00
11/8/2017 13:00
11/28/2017 15:00

Note that the data and time records are separated by a space.

Task: Separate the single DATETIME variable into two separate columns (variables) as date and time

SAS Code (E466)

```
LIBNAME any 'c:\sas_exercise';
DATA any.anyspace2;
INFILE 'C:\SAS_EXERCISE\INDEX.CSV' DLM="," DSD;
INPUT DATETIME $21.;
Myblank=ANYSPACE(Datetime);
myDate=SUBSTR(DATETIME,1,Myblank);
myTime=SUBSTR(DATETIME, Myblank+1);
RUN:

PROC PRINT DATA=any.anyspace2;
RUN;
```

LIBNAME any "c:\sas_exercise" defines a library named any representing the folder of c:\sas_exercise.

DATA any.anyspace2 creates a new dataset called anyspace2.sas7bdat and stores it in the any library.

INFILE "c:\SAS_EXERCISE\INDEX.csv" DLM="," DSD tells SAS that the INDEX.csv file in the folder of c:\sas_exercise is used to create the anyspace2.sas7bdat SAS dataset.

DLM="," tells SAS that the delimiter used in the INDEX.csv file is the , comma.

DSD further instructs SAS to handle three things:

1: Treat two consecutive delimiters (, ,) as a missing value.

2: Remove quotation marks from character values and treat everything within a quotation mark together as a single observation for the variable (e.g., "AB, C", XYZ will be treated as 2 variables, not 3. The "AB, C" is treated as 1).

3: Assume data separator is the , comma.

INPUT DATETIME $21. tells SAS that the variable DATETIME in the INDEX.csv file is a character variable with a length of 21 columns.

Myblank= ANYSPACE(Datetime) creates a new variable named Myblank by using SAS function ANYSPACE. The ANYSPACE(Datetime) specification determines the column number where a blank first appears in the Datetime observation.

myDate=SUBSTR(DATETIME,1,Myblank) creates a new variable called myDate through SAS function of SUBSTR. The SUBSTR function extracts 1st column and all the way to the column where the blank (Myblank) starts.

myTime=SUBSTR(DATETIME, Myblank+1) creates a new variable called myTime through SAS function of SUBSTR. The SUBSTR function extraction starts at column 1 after the blank column (Myblank) and retains all columns to the right.

RUN kicks off the DATA step.

PROC PRINT DATA=any.anyspace2 directs SAS to print the dataset anyspace2.sas7bdat in the any library on the screen.

RUN kicks off the PROC PRINT step.

Obs	DATETIME	DATETIME3	Myblank	myDate	myTime
1	1/1/2017 1:00	1/1/2017 1:00	9	1/1/2017	1:00
2	1/1/2017 20:00	1/1/2017 20:00	9	1/1/2017	20:00
3	1/12/2017 23:00	1/12/2017 23:00	10	1/12/2017	23:00
4	11/8/2017 13:00	11/8/2017 13:00	10	11/8/2017	13:00
5	11/28/2017 15:00	11/28/2017 15:00	11	11/28/2017	15:00

In this example, the DATETIME variable's format has no standard DATETIME **INFORMAT** which can be used (of course, you can define your own INFORMAT. But that's not the point here). By splitting the column (DATETIME variable) into date and time, you read the DATETIME data correctly.

5

Basic Data Analysis

5.1 USING THE BY STATEMENT TO GROUP AND SORT DATA

During data analysis, grouping and sorting data by attributes such as age, gender, day of the week, and month of the year. are routine practice. For certain SAS procedures, you will need to sort your data first before such procedures can be used.

Example: 5.1

Data: Truck.csv data is listed below.

ID	AreaType	RoadwayType	TruckPercent
CA356	Rural	Interstate	0.071
PK567	Rural	Secondary	0.098
CA781	Urban	Interstate	0.122
CA665	Urban	Interstate	0.157
DC236	Rural	Interstate	0.083
DC672	Rural	Secondary	0.025
DC123	Rural	Secondary	0.185
DC341	Urban	Interstate	0.114
VA665	Urban	Secondary	0.189

Task: Sort the data by

1. AreaType
2. RoadwayType
3. RoadwayType first and then AreaType.

SAS Codes (E51)

```
DATA mytruck  ;
FORMAT RoadwayType $10.;
INFILE  'c:\sas_exercise\truck5p1.csv' DLM=','
FIRSTOBS=2  DSD MISSOVER;
INPUT ID $ AreaType $ RoadwayType $ TruckPercent ;
```

```
RUN;

PROC SORT DATA=mytruck OUT=byAT;
BY AreaTYPE;
RUN;
PROC PRINT DATA=byAT (Obs=6);
RUN;
```

DATA mytruck creates a new SAS dataset called mytruck.sas7bdat and stores it in the default WORK library.

FORMAT RoadwayType $10. instructs SAS to display the RoadwayType as a 10-column wide character variable.

INFILE tells SAS that the file to be read is called truck5p1.csv stored in the c:\sas_exercise folder.

DSD further instructs SAS to handle three things:
1: Treat two consecutive delimiters (, ,) as a missing value.
2: Remove quotation marks from character values and treat everything within a quotation mark together as a single observation for the variable (e.g., "AB, C", XYZ will be treated as 2 variables, not 3. The "AB, C" is treated as 1).
3: Assume data separator is the , comma.

MISSOVER prevents SAS from going to the next data line if it can't locate values with the current dateline for all variables listed in the INPUT statement. SAS assigns a missing value for variables without values.

INPUT statement lists all variables and their INFORMATs where ID, AreaType RoadwayType are character ($) variables and TruckPercent is a numerical variable.

RUN kicks off the DATA step.

PROC SORT invokes the Sort function. DATA to be sorted is the mytruck.sas7bdat. The sorting result is to be stored in a file named byAT in the WORK library as specified by the OUT= byAT option.

Sorting is by AreaTYPE in the default ascending order (A to Z). You can specify the result be ordered by descending sequence.

RUN kicks off the PROC SORT procedure.

PROC PRINT DATA=byAT (Obs=6) directs SAS to print the first 6 rows of the result on the screen for the data named ByAT.sas7bdat.

RUN kicks off the PROC PRINT procedure.

Result:

Obs	RoadwayType	ID	AreaType	TruckPercent
1	Interstate	DC236	Rural	0.083
2	Secondary	DC672	Rural	0.025
3	Secondary	DC123	Rural	0.185
4	Secondary	VA345	Rural	0.128
5	Interstate	PK123	Rural	0.095
6	Interstate	CA788	Rural	0.153

```
PROC SORT DATA=mytruck OUT=byRT;
BY RoadwayTYPE;
RUN;
PROC PRINT DATA=byRT (Obs=7);
RUN;
```

PROC SORT invokes the Sort function. DATA to be sorted is the mytruck.sas7bdat. The sorting result is to be stored in a file named byRT in the WORK library as specified by the OUT= byRT option.

Sorting is by RoadwayTYPE in the default ascending order (A to Z). You can specify the result to be ordered by descending sequence.

RUN kicks off the PROC SORT procedure.

PROC PRINT DATA=byRT (Obs=7) directs SAS to print the first 7 rows of the result on the screen for the data named byRT.sas7bdat.

RUN kicks off the PROC PRINT procedure.

Result:

Obs	RoadwayType	ID	AreaType	TruckPercent
1	Interstate	CA356	Urban	0.071
2	Interstate	CA781	Urban	0.122
3	Interstate	CA665	Urban	0.157
4	Interstate	DC236	Rural	0.083
5	Interstate	DC341	Urban	0.114
6	Interstate	PK123	Rural	0.095
7	Interstate	CA788	Rural	0.153

```
PROC SORT DATA=mytruck OUT=byRT_AT;
BY RoadwayTYPE AreaType;
RUN;

PROC PRINT DATA=byRT_AT (Obs=8);
RUN;
```

PROC SORT invokes the Sort function. Data to be sorted is the mytruck.sas7bdat. The sorting result is to be stored in a file named byRT_AT in the WORK library as specified by the OUT= byRT_AT option.

Sorting is by RoadwayType first and then by AreaType in ascending order (A to Z). You can also specify the result to be ordered by descending sequence.

RUN kicks off the PROC SORT procedure.

PROC PRINT DATA=byRT_AT (Obs=8) directs SAS to print the first 8 rows of the result on the screen for the data named byRT_AT.sas7bdat.

RUN kicks off the PROC PRINT procedure.

Result:

Obs	RoadwayType	ID	AreaType	TruckPercent
1	Interstate	DC236	Rural	0.083
2	Interstate	PK123	Rural	0.095
3	Interstate	CA788	Rural	0.153
4	Interstate	CK341	Rural	0.198
5	Interstate	CA356	Urban	0.071
6	Interstate	CA781	Urban	0.122
7	Interstate	CA665	Urban	0.157
8	Interstate	DC341	Urban	0.114

5.2 UNDERSTANDING FIRST.variable AND LAST.variable

When the **SET** and **BY** statements are used together, the combination automatically triggers the creation of two temporary variables: **FIRST.**variable and **LAST.**variable. Before you use the **SET** and **BY** statements, your data must be sorted by the variable used in the **BY** and **SET** statements.

When a dataset is sorted, its rows are ordered alphabetically based on the **BY** variable. You can specify if you want the sequence ascending or descending by using the sort option statement.

For rows containing observations belonging to a subgroup defined by the **BY** variable, these rows are ordered by their order in their original dataset.

For example, you have 100 observations under a variable named "Vehicle." The "Vehicle" has three different groups: truck, motorcycle, and car. In the original unsorted file, thirty (30) rows are trucks that are scattered throughout the 100-row dataset. Ten (10) rows are motorcycles that are scattered throughout the 100-row dataset. And lastly, sixty (60) rows are cars that are also scattered throughout the 100-row dataset.

When you sort the above data with the statement "**BY** Vehicle" in ascending order, a newly sequenced dataset based on the order of the observations under "car," "motorcycle," and "truck" is generated.

In this newly sequenced dataset, 60 car rows are listed first, followed by the 10 motorcycle rows, and lastly, the 30 truck rows. The order is cars, motorcycles, and truck because you specified an ascending order.

Within the first 60 car rows, the first row is the first car row (observation) that appeared in the original 100-row dataset (e.g., it may be the 21st row in the original 100-row dataset). The 60th car row is the last car observation (row) that appeared in the original 100-row dataset. This ordering sequence repeats itself for the subgroup of motorcycle and truck.

For the sorted dataset, the **FIRST**.variable function assigns the variable a value of 1 to the first-row observation in a **BY** group and the value of 0 for all other row observations in the **BY** group. The **LAST**.variable function assigns the variable a value of 1 to the last observation in a **BY** group and the value of 0 for all other row observations in the **BY** group.

With the above example, the first row for the FIRST.variable of the 60 car rows will be assigned a 1. The other 59 car rows will be assigned 0. From the first observation in the car row to the 59th car row, the LAST.variable will be assigned 0. The 60th observation, which is the last car row, is assigned 1.

Example: 5.2

Data: mydata.sas7bdat
The mydata.sas7bdat has three variables: Highway ID, Direction, and Number of Vehicles.

HW_id	Direction	Vehicles
66	E	1321
17	S	2178
19	N	3215
1	N	3215
8	E	1785
38	E	1265
66	W	1200
1	N	1254
27	W	4565
66	E	1256
1	N	3256
27	S	1567

Task: Use the **SET** and **BY** statements to illustrate the **FIRST**.var and **LAST**.var

SAS Code (E52a)

```
LIBNAME yourdata 'c:\sas_exercise';
DATA mydata2;
SET yourdata.mydata;
RUN;
PROC SORT DATA=mydata2
OUT=mydata2_sorted;
BY HW_ID;
RUN;
PROC PRINT data=mydata2_sorted;
RUN;
```

LIBNAME yourdata "c:\sas_exercise" defines a library named yourdata representing the folder of c:\sas_exercise.

DATA mydata2 creates a new dataset called mydata2 in the WORK library.

SET yourdata.mydata tells SAS that mydata.sas7bdat in the yourdata library is used to create the mydata2.sas7bdat.

RUN kicks off the DATA step.

Sorting Result:

Obs	HW_id	direction	vehicles
1	1	N	3215
2	1	N	1254
3	1	N	3256
4	8	ES	1785
5	17	S	2178
6	19	N	3215
7	27	W	4565
8	27	S	1567
9	38	ES	1265
10	66	E	1321
11	66	W	1200
12	66	ES	1256

PROC SORT DATA=mydata2 OUT=mydata2_sorted invokes the SORT function to sort mydata2. sas7bdat in the WORK library and output the result as file mydata2_ sorted.sas7bdat in the WORK library. The sorting is based on HW_ID (BY HW_ID) in the default ascending order.

RUN kicks off the PROC SORT procedure.

PROC PRINT data=mydata2_sorted instructs SAS to put the mydata2_ sorted.sas7bdat in the WORK library on the screen.

RUN kicks off the PROC PRINT procedure.

SAS Code (E52b)

```
LIBNAME yourdata 'c:\sas_exercise';
DATA mydata2;
SET yourdata.mydata;
RUN;

PROC SORT DATA=mydata2
OUT=mydata2_sorted;
BY HW_ID;
RUN;

DATA my_sorted;
SET mydata2_sorted;
BY  HW_ID;
My_First_ID= FIRST.HW_ID;
My_Last_ID= LAST.HW_ID;
RUN;

PROC PRINT data=my_sorted;
RUN;
```

LIBNAME yourdata "c:\sas_exercise" defines a library named yourdata representing the folder of c:\sas_exercise.

DATA mydata2 creates a new dataset called mydata2 in the WORK library.

SET yourdata.mydata tells SAS that mydata.sas7bdat in the yourdata library is used to create the mydata2. sas7bdat.

RUN kicks off the DATA step.

PROC SORT DATA=mydata2 OUT=mydata2_sorted invokes the SORT function to sort mydata2. sas7bdat in the WORK Library and output the result as mydata2_sorted. sas7bdat to the WORK library. The sorting is based on HW_ID (BY HW_ ID) in the default ascending order.

RUN kicks off the PROC SORT procedure.

Result:

Obs	HW_id	direction	vehicles	My_First_ID	My_Last_ID
1	1	N	3215	1	0
2	1	N	1254	0	0
3	1	N	3256	0	1
4	8	ES	1785	1	1
5	17	S	2178	1	1
6	19	N	3215	1	1
7	27	W	4565	1	0
8	27	S	1567	0	1
9	38	ES	1265	1	1
10	66	E	1321	1	0
11	66	W	1200	0	0
12	66	ES	1256	0	1

DATA my_sorted creates a new SAS dataset named my_sorted and stores it in the WORK library.

SET mydata2_sorted tells SAS to read mydata2_sorted in the WORK library.

The BY HW_ID statement used in combination with the SET statement triggers the automatic generation of two temporary variables called FIRST.HW_ID and LAST.HW_ID to characterize row order and appearance.

My_First_ID=FIRST.HW_ID assigns the auto-generated FIRST.HW_ID to a new variable named My_First_ID, which becomes part of the my_sorted.sas7bdat file.

My_Last_ID=LAST.HW_ID assigns the auto-generated LAST.HW_ID to a new variable named My_Last_ID, which becomes part of the my_sorted.sas7bdat file.

PROC PRINT data=my_sorted instructs SAS to print the data my_sorted.sas7bdat to the screen.

RUN kicks off the PROC PRINT procedure.

5.3 REMOVING DUPLICATE RECORDS (ROWS)

When dealing with datasets, you should always check for duplicate records (rows). There are wide specifications in what constitutes a duplicate record. A duplicate could require all observations to be identical across multiple rows (identical rows), or just one observation for one variable to be repeated among multiple rows (e.g., multiple identical last names from different rows).

Be very cautious when you remove duplicate records. Always keep a copy of the original file. Consult with the data owner or the subject matter expert when in doubt.

5.3.1 Repeated (Identical) Rows – NODUP

If your criteria for duplicate records are that observations among different rows need to be identical (identical rows), you can use the **NODUP** option as illustrated below.

You must sort your dataset first by using the **BY _ALL_** statement because the **NODUP** only compares rows adjacent to each other.

Example: 5.3.1

Data: dup-original.sas7bdat
In the example data below, rows 3, 4, and 8 are identical rows. You can use the **NODUP** command with "**PROC SORT; BY _ALL_;**" statements to remove duplicates. However, if you do not use the " **BY _ALL_;**" statement, only row # 4 will be removed. Row # 8 will stay because it is not next to either row 3 or 4. The **NODUP** command, as stated earlier, only checks rows adjacent to each other.

Obs	RoadID	Ownership	Length	Lanes	Condition	Truck
1	US26	Talbert	12.5	6	g	y
2	SR24	Talbert	6.3	4	a	y
3	CR24	Leon	2.4	4	g	y
4	CR24	Leon	2.4	4	g	y
5	CR24	Flagler	2.5	4	g	y
6	US124	Santos	26.7	4	a	y
7	CR24	Beach	12.1	2	u	y
8	CR24	Leon	2.4	4	g	y
9	US26	Talbert	12.6	6	g	y
10	US88	Park	2.2	4	e	n

Task: Remove identical rows

SAS Code (E531) – with sort

```
LIBNAME myd 'c:\sas_exercise';
DATA Exer1;
SET myd.dup_original;
PROC SORT DATA=Exer1 NODUP OUT=myd.dup_exer1;
BY _ALL_;
RUN;
PROC PRINT DATA=myd.dup_exer1;
RUN;
```

LIBNAME myd "c:\sas_exercise" defines a library named myd representing the folder of c:\ sas_exercise.

DATA Exer1 creates a new SAS dataset called Exer1.sas7bdat in the WORK library.

SET myd.dup_original tells SAS that a data file named dup_original.sas7bdat in the myd library is to be read for the creation of the Exer1.sas7bdat file.

PROC SORT Data=Exer1 NODUP instructs SAS to sort the data Exer1.sas7bdat where the NODUP further instructs SAS to delete all identical rows (based on the BY variable specification) (keep only one row). The final output (OUT=myd.dup_exer1) is to output the final file as dup_exer1 in the myd library (a permanent file is created given it is not stored in the WORK library).

BY _ALL_ is part of the PROC SORT command. The sorting is to be done for all variables meaning column 1 gets sorted first, column 2 gets sorted second, column 3, etc. until all columns are sorted.

RUN kicks off the PROC SORT procedure.

PROC PRINT DATA=myd.dup_exer1 instructs SAS to print out the dup_exer1.sas7bdat in the myd library.

RUN kicks off the PROC PRINT procedure.

Result:

Obs	RoadID	Ownership	Length	Lanes	Condition	Truck
1	CR24	Beach	12.1	2	u	y
2	CR24	Flagler	2.5	4	g	y
3	CR24	Leon	2.4	4	g	y
4	SR24	Talbert	6.3	4	a	y
5	US124	Santos	26.7	4	a	y
6	US26	Talbert	12.5	6	g	y
7	US26	Talbert	12.6	6	g	y
8	US88	Park	2.2	4	e	n

5.3.2 Customized Duplicate Criteria – NODUPKEY

Your dataset may have many variables. You can specify which variable observations if repeated, are the basis for defining duplication using the **NODUPKEY** option. The **NODUPKEY** option eliminates observations that share a **BY** variable.

Your "**By** variable" could be just one variable (e.g., Last Name) or all variables (e.g., Last Name, First Name, DOB, Gender, Birth State).

Be very careful when using the **NODUPKEY** option. Unlike NODUP, where a row is eliminated only when all observations are the same, NODUPKEY will remove a row solely based on the repeat of the "**BY** variable."

Example: 5.3.2

Data: dup.original.sas7bdat

RoadID	Ownership	Length	Lanes	Condition	Truck
US26	Talbert	12.5	6	g	y
SR24	Talbert	6.3	4	a	y
CR24	Leon	2.4	4	g	y
CR24	Leon	2.4	4	g	y
CR24	Flagler	2.5	4	g	y
US124	Santos	26.7	4	a	y
CR24	Beach	12.1	2	u	y
CR24	Leon	2.4	4	g	y
US26	Talbert	12.6	6	g	y
US88	Park	2.2	4	e	n

Task: Remove duplicate records based on RoadID and Ownership. The criteria are that when both RoadID and Ownership repeat among rows, these rows are considered duplicates.

SAS Codes (E532)

```
LIBNAME myd 'c:\sas_exercise';
DATA d2;
SET myd.dup_original;
PROC SORT DATA=d2 NODUPKEY OUT=myd.dup_original2a;
BY RoadID Ownership;
RUN;

PROC PRINT DATA=myd.dup_original2a;
RUN;
```

LIBNAME defines a library named myd representing the folder of c:\sas_exercise.

DATA creates a new SAS dataset called d2.sas7bdat in the WORK library.

SET tells SAS that a data file named dup_original.sas7bdat in the myd library is to be read for the creation of the d2.sas7bdat file.

PROC SORT DATA=d2 NODUPKEY instructs SAS to sort the data d2.sas7bdat where the NODUPKEY further instructs SAS to delete all identical rows (based on the BY variable specification). The final output (OUT=myd.dup_original2a) is in the myd library with a file named dup_original2a.sas7bdat (a permanent file is created given it is not stored in the WORK library).

BY RoadID Ownership is part of the PROC SORT command. It sorts the data based on RoadID first and then Ownership.

Rows repeat themselves with similar RoadID and Ownership (both are considered identical (duplicate)).

RUN kicks off the PROC SORT procedure.

PROC PRINT DATA=myd.dup_original2a instructs SAS to print out the dup_original2a.sas7bdat in the myd library.

RUN kicks off the PROC PRINT procedure.

Result:

Obs	RoadID	Ownership	Length	Lanes	Condition	Truck
1	CR24	Beach	12.1	2	u	y
2	CR24	Flagler	2.5	4	g	y
3	CR24	Leon	2.4	4	g	y
4	SR24	Talbert	6.3	4	a	y
5	US124	Santos	26.7	4	a	y
6	US26	Talbert	12.5	6	g	y
7	US88	Park	2.2	4	e	n

5.3.3 Using FIRST.var and LAST.var for Duplicate Identification

You can use the **SET** and **BY** statements to generate the **FIRST**.variable and the **LAST**.variable. You can then use the **FIRST**.variable and **LAST**.variable for duplicate identification.

For a given observation (row), if both its **FIRST**.var=1 and its **LAST**.var=1, it indicates that the row is a single unique row (observation).

When a row's **FIRST**.var=0 and **LAST**.var=0, it means it is a repeated row and suggests the possible existence of additional repeated rows.

When a row's **FIRST**.var=0 and **LAST**.var=1, it means that the row is the last repeated row based on the criteria specified by the BY variable.

Example: 5.3.3

Data: dup_original.sas7bdat

RoadID	Ownership	Length	Lanes	Condition	Truck
US26	Talbert	12.5	6	g	y
SR24	Talbert	6.3	4	a	y
CR24	Leon	2.4	4	g	y
CR24	Leon	2.4	4	g	y
CR24	Flagler	2.5	4	g	y
US124	Santos	26.7	4	a	y
CR24	Beach	12.1	2	u	y
CR24	Leon	2.4	4	g	y
US26	Talbert	12.6	6	g	y
US88	Park	2.2	4	e	n

Task: Remove duplicate records and generate new datasets

SAS Codes (E533)

```
LIBNAME myd 'c:\sas_exercise';
```

```
PROC SORT DATA=myd.dup_original
out=my_sorted1;
BY RoadID;
RUN;

DATA my_sorted2;
SET my_sorted1;
BY  RoadID;
My_First_RoadID= FIRST.RoadID;
My_Last_RoadID= LAST.RoadID;
RUN;

DATA Duplicate_removed_incl_Dup;
SET my_sorted2;
IF My_First_RoadID=1 and My_
Last_RoadID=1 THEN OUTPUT
Duplicate_removed_incl_Dup;
RUN;

DATA Duplicate_removed;
SET my_sorted2;
IF My_First_RoadID=1 THEN OUTPUT
Duplicate_removed;
RUN;


PROC PRINT DATA=my_sorted2;
PROC PRINT Data=Duplicate_
removed_incl_Dup ;
PROC PRINT DATA= Duplicate_
removed ;
RUN;
```

LIBNAME defines a library named myd representing the folder of c:\sas_exercise.

PROC SORT DATA=myd.dup_original invokes SAS's sort procedure to sort the dup_original.sas7bdat file in the myd library by the variable RoadID (BY RoadID). The result is put in the WORK folder in a file name my_sorted1 (out=my_sorted1).

DATA my_sorted2 creates a new SAS dataset called my_sorted2.sas7bdat in the WORK library.

SET my_sorted1 tells SAS that a data file named my_sorted1.sas7bdat in the WORK library is to be read and used for the creation of my_sorted.sas7bdat file.

BY RoadID is used with the SET my_sorted1 statement together to trigger the autogeneration of temporary FIRST.RoadID and LAST.RoadID variables.

My_First_RoadID=FIRST.RoadID and My_Last_RoadID=LAST.RoadID create two new permanent variables named My_First_RoadID and My_Last_RoadID, which are part of the my_sorted2.sas7bdat file.

RUN kicks off the DATA step.

DATA creates a new dataset named Duplicate_removed_incl_Dup.sas7bdat in the WORK library.

SET my_sorted2 tells SAS to use my_sorted2.sas7bdat in the WORK library to generate the Duplicate_removed_incl_Dup.sas7bdat file.

The file to be generated needs to meet the condition of "IF My_First_RoadID=1 and My_Last_RoadID=1." The resulting data Duplicate_removed_incl_Dup is delivered to the WORK library per the OUTPUT command.

Run kicks off the DATA step.

DATA creates a new SAS dataset called Duplicate_removed.sas7bdat.

SET my_sorted2 tells SAS to use the my_sorted2.sas7bdat file in the WORK library to create the Duplicate_removed data based on the condition prescribed in the IF My_First_RoadID=1.

Observations (rows) meeting the condition are sent to the Duplicate_removed.sas7bdat file in the WORK library (OUTPUT Duplicate_removed).

Run kicks off the DATA step.

The listed condition can be met by any first appearing unique rows (MY_First_RoadID=1). The three PROC PRINT statements print all the data in the WORK library on the screen. RUN kicks off the PROC PRINT procedure.

Result:

my_sorted2.sas7bdat

Obs	RoadID	Ownership	Length	Lanes	Condition	Truck	My_First_RoadID	My_Last_RoadID
1	CR24	Leon	2.4	4	g	y	1	0
2	CR24	Leon	2.4	4	g	y	0	0
3	CR24	Flagler	2.5	4	g	y	0	0
4	CR24	Beach	12.1	2	u	y	0	0
5	CR24	Leon	2.4	4	g	y	0	1
6	SR24	Talbert	6.3	4	a	y	1	1
7	US124	Santos	26.7	4	a	y	1	1
8	US26	Talbert	12.5	6	g	y	1	0
9	US26	Talbert	12.6	6	g	y	0	1
10	US88	Park	2.2	4	e	n	1	1

`Duplicate_removed_incl_Dup.sas7bdat`

Obs	RoadID	Ownership	Length	Lanes	Condition	Truck	My_First_RoadID	My_Last_RoadID
1	SR24	Talbert	6.3	4	a	y	1	1
2	US124	Santos	26.7	4	a	y	1	1
3	US88	Park	2.2	4	e	n	1	1

`Duplicate_removed.sas7bdat`

Obs	RoadID	Ownership	Length	Lanes	Condition	Truck	My_First_RoadID	My_Last_RoadID
1	CR24	Leon	2.4	4	g	y	1	0
2	SR24	Talbert	6.3	4	a	y	1	1
3	US124	Santos	26.7	4	a	y	1	1
4	US26	Talbert	12.5	6	g	y	1	0
5	US88	Park	2.2	4	e	n	1	1

5.4 USING FIRST.variable AND LAST.variable TO COUNT AND QUANTIFY

As we have learned, when using **SET** and **BY** statements together, SAS automatically creates two temporary variables: **FIRST**.var and LAST.var. You can use the **FIRST**.var and **LAST**.var to count the number of observations or accumulations of observations (e.g., length, time, weight) in a group.

Example: 5.4

Data: speed_c.sas7bdat
The speed data has a total of 65,537 observations (rows) as illustrated below.

The data contain the **speed** (mph) at a given time (**Mtime**) at a given date (**Mdate**) for a specific roadway segment (**Segment ID**). Additionally, the data contain information regarding **length**, **type**, and **ownership**.

SegmentID	owner	type	length	Mdate	Mtime	speed
PG8912	state	1	0.25	1-Jan-17	0:00:00	54
PG8912	state	1	0.25	1-Jan-17	1:00:00	58
PG8912	state	1	0.25	1-Jan-17	2:00:00	57
PG8912	state	1	0.25	1-Jan-17	3:00:00	59
PG8912	state	1	0.25	1-Jan-17	4:00:00	57
PG8912	state	1	0.25	1-Jan-17	5:00:00	61
PG8912	state	1	0.25	1-Jan-17	6:00:00	54

Task:

1. Create a new variable "Congestion" for each road segment based on the speed criteria of **Red** =< 25 mph, 25 mph< **Yellow** =<40 mph, and **Green** >40 mph.
2. Determine the number of segments (counts) and cumulative segment lengths (miles) by owner (city, county, state).
3. Determine roadway length by the "Congestion" status.

SAS Codes for Task 1: create the new variable "Congestion"

SAS Codes (E54t1)

```
LIBNAME sp 'c:\sas_exercise';
DATA Conge;
FORMAT Congestion $6.;
SET sp.speed_c;
IF Speed=<25 THEN Congestion='Red';
ELSE IF 25<speed =<40 THEN Congestion='Yellow';
ELSE IF speed>40 THEN Congestion ='Green';
RUN;
```

```
PROC PRINT DATA=Conge (OBS=8);
RUN;
```

LIBNAME sp "c:\sas_exercise" defines a library named sp representing the folder of c:\sas_exercise.

DATA Conge creates a new SAS data called Conge.sas7bdat in the WORK library.

FORMAT Congestion $6. defines Congestion as a character variable with 6-column width (keep in mind that FORMAT affects how observations are displayed and not the accuracy. INFORMAT affects how observations are read and their accuracy).

SET sp.speed_c instructs SAS to use the file speed_condition_2.sas7bdat in the sp library to create the Conge.sas7bdat file.

The IF Speed=<25 THEN Congestion="Red" statement creates a new variable named Congestion, and it fills the Congestion observation with Red if the Speed equals to or is less than 25.

Statements of ELSE IF … continue all other Congestion conditions.

RUN kicks off the DATA step.

PROC PRINT DATA=Conge (OBS=8) instructs SAS to print the first 8 rows of Conge.sas7bdat in the default WORK library.

RUN kicks off the PROC PRINT procedure.

Task 1 Result: the new variable Congestion is created

Obs	Congestion	SegmentID	owner	type	length	Mdate	Mtime	speed
1	Green	PG8912	state	1	0.25	01JAN2017	0:00:00	54
2	Green	PG8912	state	1	0.25	01JAN2017	1:00:00	58
3	Green	PG8912	state	1	0.25	01JAN2017	2:00:00	57
4	Green	PG8912	state	1	0.25	01JAN2017	3:00:00	59
5	Green	PG8912	state	1	0.25	01JAN2017	4:00:00	57
6	Green	PG8912	state	1	0.25	01JAN2017	5:00:00	61
7	Green	PG8912	state	1	0.25	01JAN2017	6:00:00	54
8	Green	PG8912	state	1	0.25	01JAN2017	7:00:00	60

SAS Codes for Task 2a: Counts of segments by owner

SAS Codes (E54t2a)

```
LIBNAME my 'c:\sas_exercise';

PROC SORT DATA=my.speed_c out=speed_s;
BY owner ;
RUN;
```

```
DATA countBYowner;
SET speed_s ;
BY owner ;
IF FIRST.owner THEN
SegmentCount=0;
SegmentCount+1;
IF LAST.owner;
RUN;

PROC PRINT
data=countBYowner noobs;
FORMAT SegmentCount
comma6. ;
var owner SegmentCount;
RUN;
```

Task 2a Result: Counts of segments by owner

owner	SegmentCount
city	13,065
county	26,212
state	26,260

LIBNAME my "c:\sas_exercise" defines a my library representing the folder of c:\sas_exercise.

PROC SORT Data=my.speed_c invokes the sort function to sort the speed_c.sas7bdat file in the my library BY owner in ascending (default) order. The sorted output file is named speed_s.sas7bdat stored in the WORK library.

RUN kicks off the PROC SORT procedure.

DATA creates a new SAS dataset named countBYowner.sas7bdat in the WORK library.

SET instructs SAS to use the speed_s.sas7bdat file in the WORK library for the creation of countBYowner.sas7bdat file.

By owner in combined usage with the SET command triggers the autogeneration of the two temporary FIRST.owner and LAST.owner variables.

As SAS reads observations row by row, for a given owner (after sorting by owner), if it is the first time the row is read, then the FIRST.owner=1 (IF FIRST.owner). And a new variable called SegmentCount is set to 0 and a counter SegmentCount+1 is set. When row 1 is read, the variable SegmentCount=0+1. This counting continues until the LAST.owner equaling to 1 row (IF LAST.owner) (same as IF LAST.owner=1) is read. This means, for the current owner, all rows have been counted. The counter will be reset to 0 (SegmentCount=0) for the next owner as SAS reads the next owner data row by row.

PROC PRINT data=countBYowner noobs instructs SAS to print the data of countBYowner.sas7bdat without displaying any data (noobs) until further instruction.

FORMAT SegmentCount comma6. instructs SAS to display the SegmentCount with a 6-column integer number separated with the thousands mark.

var owner SegmentCount is the further specification for the PROC PRINT command. Now SAS will only print out variables owner and SegmentCount on the screen.

Run kicks off the PROC PRINT procedure.

SAS Codes for Task 2b: Segment length by owner

SAS Codes (E54t2b)

```
LIBNAME sp2 'c:\sas_exercise';
PROC SORT DATA=sp2.speed_c out=speed_s;
BY owner ;
```

```
RUN;

DATA lengthBYowner;
SET speed_s ;
BY owner ;
IF FIRST.owner THEN CumSegLength=0;
CumSegLength+length;
IF LAST.owner;
RUN;
PROC PRINT DATA=lengthBYowner noobs;
FORMAT CumSegLength comma6. ;
var owner CumSegLength;
RUN;
```

LIBNAME sp2 "c:\sas_exercise" defines an sp2 library representing the folder of c:\sas_exercise.

PROC SORT Data=sp2.speed_c invokes the sort function to sort the speed_c.sas7bdat file in the sp2 library BY owner in ascending (default) order. The sorted output file is named speed_s.sas7bdat stored in the WORK library.

RUN kicks off the PROC SORT procedure.

DATA creates a new SAS dataset named lengthBYowner.sas7bdat in the WORK library.

SET speed_s instructs SAS to use the speed_s.sas7bdat file in the WORK library for the creation of lengthBYowner.sas7bdat file.

BY Owner in combined usage with the SET command triggers the autogeneration of the two temporary FIRST.owner and LAST.owner variables.

As SAS reads observations row by row, for a given owner (after sorting by owner), if it is the first time the row is read, then the FIRST.owner=1 (IF FIRST.owner); And a new variable called CumSegLength is set to 0 and an accumulator CumSegLength+length is set. When row 1 is read, the variable CumSegLength=0+length. This addition continues until the LAST.owner equaling to 1 row (IF LAST.owner) (same as IF LAST.owner=1) is read. This means, for the current owner, all rows have been counted. The counter will be reset to 0 (CumSegLength=0) for the next owner as SAS reads the next owner.

PROC PRINT data=lengthBYowner noobs instructs SAS to print the data of lengthBYOwner.sas7bdat without displaying any data (noobs) until further instruction.

FORMAT CumSegLength comma6. instructs SAS to display the CumSegLength with a 6-column integer number separated with the thousands mark.

var owner CumSegLength is the further specification for the PROC PRINT command. Now SAS will only print out variables owner and CumSegLength on the screen.

Run kicks off the PROC PRINT procedure.

Task 2b Result: Roadway segment length by owner

owner	CumSegLength
city	11,489
county	10,489
state	14,011

SAS Codes for Task 3: Segment length and count by congestion status

SAS Codes (E54t3)

```
LIBNAME sp 'c:\
sas_exercise';
DATA C;
FORMAT Congestion $6.;
SET
sp.speed_condition_2;
IF Speed=<25 THEN
Congestion='Red';
ELSE IF 25<speed =<40
THEN
Congestion='Yellow';
ELSE IF speed>40 THEN
Congestion ='Green';
RUN;

PROC SORT DATA=C;
BY Congestion ;
RUN;

DATA task3;
SET C ;
BY Congestion ;
IF FIRST.congestion THEN
CumSegLength=0 and
CumSegC=0;
CumSegLength+length;
CumSegC+1;
IF LAST.congestion;
RUN;

PROC PRINT DATA=task3
noobs;
FORMAT CumSegLength
CumSegC comma6. ;
var Congestion
CumSegLength  CumSegC;
RUN;
```

LIBNAME sp "c:\sas_exercise" defines a library named sp representing the folder of c:\sas_exercise.

DATA C creates a new SAS dataset called C and stores it in the WORK folder.

FORMAT Congestion $6. instructs SAS to display Congestion as a 6-column width character variable.

SET sp.speed_condition_2 tells SAS to use the speed_condition_2.sas7bdat file in the sp library to create the C dataset.

IF Speed =< 25 THEN Congestion="Red" creates a new variable called Congestion and when the speed is<or=25, the observation for Congestion is filled with Red.

ELSE IF … THEN continues the speed conditions for different Congestion status (yellow, green).

RUN kicks off the DATA step.

PROC SORT DATA=C and BY Congestion tell SAS to sort C.sas7bdat in the WORK library by the Congestion status in the default ascending order.

RUN kicks start the PROC SORT procedure.

DATA task3 creates a new SAS dataset named task3 stored in the default WORK library.

Set C tells SAS to use the C.sas7bdat in the WORK library to create the taks3.sas7bdat file based on the sorted result of BY Congestion.

When SET and BY Congestion used together, it triggers the autogeneration of two temporary variables FIRST.congestion and LAST.congestion.

IF FIRST.congestion THEN CumSegLength=0 and CumSegC=0 create 2 new variables called CumSegLength and CumSegC, where both of them are set to 0 if the row of observation is the 1st row of data SAS reads in the sorted file of task3 as instructed by the IF FIRST.congestion (same as IF FIRST.congestion=1).

CumSegLength+length tells SAS to add the length to the CumSegLength to obtain the latest cumulative segment length.

CumSegC+1 tells SAS to add 1 to the CumSegC every time a row is read by SAS within the BY Congestion group.

For example, the first data row among all similar subgroups of the Congestion (e.g., Red) is read, length =2.1, CumSegLength=0, CumSegLength+length=0+2.1=2.1, and CumSegC+1=0+ 1=1.

The second RED row is read and it has length =0.3, CumSegLength= 2.1+ 0.3=2.4, and CumSegC+1= 1+1=2.

The third RED row is read by SAS and it has length=0.8, CumSegLength=2.4+0.8=3.2, and CumSegC+1=2+1=3.

If the third row is the last RED row (IF LAST.congestion) (same as IF LAST.congestion=1), then the counter-statement reset itself to 0 (IF FIRST.congestion Then CumSegLength=0 and CumSegC=0). And a new round of reading and calculation starts until all the subgroups (Yellow, Green) associated with the BY Congestion are read.

PROC PRINT data=task3 noobs instructs SAS to print the data of task3.sas7bdat without displaying any data (noobs) until further instruction.

FORMAT CumSegLength CumSegC comma6. instructs SAS to display the CumSegLength and CumSegC with a 6-column integer number separated with THE thousands mark.

var Congestion CumSegLength CumSegC are the further specification for the PROC PRINT command. Now SAS will print out these three variables on the screen.

Run kicks off the PROC PRINT procedure.

Task 3 Result: Segment length and count by congestion status

Congestion	CumSegLength	CumSegC
Green	32,550	56,857
Red	413	57,957
Yellow	3,026	65,537

5.5 USING PROC SQL FOR SUBTOTALS AND PERCENTAGES

The computing of subtotals and percentages is a regular occurrence in data analysis. The PROC SQL procedure is a very simple but highly efficient way to carry out this task. Given that the syntax for SQL is different from SAS, we will start with a brief introduction to the basic relevant terms used in SAS PROC SQL statement.

SAS refers to a data file as data. A SAS dataset has observations and variables. SQL refers a data file as a table with rows and columns. The basic syntax of PROC SQL is illustrated below.

```
DATA newdata;
SET  olddata;
RUN;

PROC SQL;
    CREATE TABLE newdata AS
    SELECT *
   FROM olddata
   WHERE expression
   GROUP BY column(s)
   ORDER BY column(s);
QUIT;
```

By using the **DATA** and **SET** statements, the above SAS program creates a new dataset called **newdata** based on the file **olddata** in the WORK library.

The SQL codes above do the same thing. It creates a table called **newdata** by selecting **all variables** (as signified by the *) from the table **olddata** in the WORK library.

GROUP BY (considered a clause in SQL) groups the data by the specified column and is similar to SAS's GROUP statement. The **ORDER BY** statement rearranges the order of variables per your specification.

Example: 5.5

Data: R4D9S50.sas7bdat
All the 50 States and the District of Columbia are grouped into 9 Divisions. The 9 Divisions are further grouped into 4 Regions. The data has 6 variables as explained below.

- Region Name
- Division Name
- State Name
- Number of Licensed Drivers in a State
- A "Yes" or "No" answer on the availability of the biannual vehicle registration program in a State.

Region	Division	State	Driver	Registration
West	Pacific	AK	579405	Yes
South	ESCentral	AL	3998767	Yes
South	WSCentral	AR	2245902	Yes
West	Mountain	AZ	5361737	Yes
West	Pacific	CA	31079970	No
West	Mountain	CO	4258174	No
Northeast	NewEngland	CT	2922797	Yes
South	Southatlantic	DC	546229	No
South	Southatlantic	DE	752072	Yes
South	Southatlantic	FL	16562355	Yes
South	Southatlantic	GA	8179943	Yes
West	Pacific	HI	1116079	Yes
Midwest	WNCentral	IA	2377524	Yes
West	Mountain	ID	1311498	Yes
Midwest	ENCentral	IL	9597855	No
Midwest	ENCentral	IN	5040991	Yes
Midwest	WNCentral	KS	2301719	Yes
South	ESCentral	KY	3474629	Yes
South	WSCentral	LA	3472615	Yes
Northeast	NewEngland	MA	5646648	No
South	Southatlantic	MD	4503033	No
Northeast	NewEngland	ME	1094948	Yes
Midwest	ENCentral	MI	7529922	Yes
Midwest	WNCentral	MN	4435973	NO
Midwest	WNCentral	MO	4914478	Yes
South	ESCentral	MS	2322325	Yes
West	Mountain	MT	794211	Yes
South	Southatlantic	NC	7737873	Yes
Midwest	WNCentral	ND	600886	Yes
Midwest	WNCentral	NE	1525202	Yes
Northeast	NewEngland	NH	1109718	Yes
Northeast	Midatlantic	NJ	7042719	No
West	Mountain	NM	1566604	Yes
West	Mountain	NV	2268602	Yes
Northeast	Midatlantic	NY	15449288	No
Midwest	ENCentral	OH	8805656	Yes
South	WSCentral	OK	2938382	Yes
West	Pacific	OR	3292643	No
Northeast	Midatlantic	PA	10124749	Yes
Northeast	NewEngland	RI	864989	Yes
South	Southatlantic	SC	3788691	Yes
Midwest	WNCentral	SD	697459	Yes
South	ESCentral	TN	5264359	No
South	WSCentral	TX	21458360	Yes
West	Mountain	UT	2483680	Yes
South	Southatlantic	VA	6347378	No
Northeast	NewEngland	VT	512375	Yes
West	Pacific	WA	2777913	No
Midwest	ENCentral	WI	4379360	Yes
South	Southatlantic	WV	1345706	Yes
West	Mountain	WY	453927	Yes

Task A:

A1) Compute a state's share (percentage) of total U.S. licensed drivers (State/∑(all States)).

Region	Division	State	Driver	Driver%
West	Pacific	AK	579405	
South	ESCentral	AL	3998767	
South	WSCentral	AR	2245902	
West	Mountain	AZ	5361737	
West	Pacific	CA	31079970	
West	Mountain	CO	4258174	
Northeast	NewEngland	CT	2922797	
South	Southatlantic	DC	546229	
South	Southatlantic	DE	752072	
South	Southatlantic	FL	16562355	
South	Southatlantic	GA	8179943	
West	Pacific	HI	1116079	
Midwest	WNCentral	IA	2377524	
West	Mountain	ID	1311498	
Midwest	ENCentral	IL	9597855	
Midwest	ENCentral	IN	5040991	
Midwest	WNCentral	KS	2301719	
South	ESCentral	KY	3474629	
South	WSCentral	LA	3472615	
Northeast	NewEngland	MA	5646648	
South	Southatlantic	MD	4503033	
Northeast	NewEngland	ME	1094948	
Midwest	ENCentral	MI	7529922	
Midwest	WNCentral	MN	4435973	
Midwest	WNCentral	MO	4914478	
South	ESCentral	MS	2322325	
West	Mountain	MT	794211	
South	Southatlantic	NC	7737873	
Midwest	WNCentral	ND	600886	
Midwest	WNCentral	NE	1525202	
Northeast	NewEngland	NH	1109718	
Northeast	Midatlantic	NJ	7042719	
West	Mountain	NM	1566604	
West	Mountain	NV	2268602	
Northeast	Midatlantic	NY	15449288	
Midwest	ENCentral	OH	8805656	
South	WSCentral	OK	2938382	
West	Pacific	OR	3292643	
Northeast	Midatlantic	PA	10124749	
Northeast	NewEngland	RI	864989	
South	Southatlantic	SC	3788691	
Midwest	WNCentral	SD	697459	
South	ESCentral	TN	5264359	
South	WSCentral	TX	21458360	
West	Mountain	UT	2483680	
South	Southatlantic	VA	6347378	
Northeast	NewEngland	VT	512375	
West	Pacific	WA	2777913	
Midwest	ENCentral	WI	4379360	
South	Southatlantic	WV	1345706	
West	Mountain	WY	453927	
	Sum(all states)			100%

A2) Compute a state's share of licensed drivers within its Region, as illustrated below (State/∑(States within the Corresponding Region)).

Region	Division	State	Driver	Driver%
Midwest	WNCentral	IA	2377524	
Midwest	ENCentral	IL	9597855	
Midwest	ENCentral	IN	5040991	
Midwest	WNCentral	KS	2301719	
Midwest	ENCentral	MI	7529922	
Midwest	WNCentral	MN	4435973	
Midwest	WNCentral	MO	4914478	
Midwest	WNCentral	ND	600886	
Midwest	WNCentral	NE	1525202	
Midwest	ENCentral	OH	8805656	
Midwest	WNCentral	SD	697459	
Midwest	ENCentral	WI	4379360	
SUM(Midwest States)				**100%**
Northeast	NewEngland	CT	2922797	
Northeast	NewEngland	MA	5646648	
Northeast	NewEngland	ME	1094948	
Northeast	NewEngland	NH	1109718	
Northeast	Midatlantic	NJ	7042719	
Northeast	Midatlantic	NY	15449288	
Northeast	Midatlantic	PA	10124749	
Northeast	NewEngland	RI	864989	
Northeast	NewEngland	VT	512375	
SUM(Northeast States)				**100%**
South	ESCentral	AL	3998767	
South	WSCentral	AR	2245902	
South	Southatlantic	DC	546229	
South	Southatlantic	DE	752072	
South	Southatlantic	FL	16562355	
South	Southatlantic	GA	8179943	
South	ESCentral	KY	3474629	
South	WSCentral	LA	3472615	
South	Southatlantic	MD	4503033	
South	ESCentral	MS	2322325	
South	Southatlantic	NC	7737873	
South	WSCentral	OK	2938382	
South	Southatlantic	SC	3788691	
South	ESCentral	TN	5264359	
South	WSCentral	TX	21458360	
South	Southatlantic	VA	6347378	
South	Southatlantic	WV	1345706	
SUM(South States)				**100%**
West	Pacific	AK	579405	
West	Mountain	AZ	5361737	
West	Pacific	CA	31079970	
West	Mountain	CO	4258174	
West	Pacific	HI	1116079	
West	Mountain	ID	1311498	
West	Mountain	MT	794211	
West	Mountain	NM	1566604	
West	Mountain	NV	2268602	
West	Pacific	OR	3292643	
West	Mountain	UT	2483680	
West	Pacific	WA	2777913	
West	Mountain	WY	453927	
SUM(West States)				**100%**

A3) Compute a state's share of licensed drivers within its Division, as illustrated below (State/∑(States within the Corresponding Division)).

Region	Division	State	Driver	Driver%
Midwest	ENCentral	IL	9597855	
Midwest	ENCentral	IN	5040991	
Midwest	ENCentral	MI	7529922	
Midwest	ENCentral	OH	8805656	
Midwest	ENCentral	WI	4379360	
sum(states)				100%
South	ESCentral	AL	3998767	
South	ESCentral	KY	3474629	
South	ESCentral	MS	2322325	
South	ESCentral	TN	5264359	
sum(states)				100%
Northeast	Midatlantic	NJ	7042719	
Northeast	Midatlantic	NY	15449288	
Northeast	Midatlantic	PA	10124749	
sum(states)				100%
West	Mountain	AZ	5361737	
West	Mountain	CO	4258174	
West	Mountain	ID	1311498	
West	Mountain	MT	794211	
West	Mountain	NM	1566604	
West	Mountain	NV	2268602	
West	Mountain	UT	2483680	
West	Mountain	WY	453927	
sum(states)				100%
Northeast	NewEngland	CT	2922797	
Northeast	NewEngland	MA	5646648	
Northeast	NewEngland	ME	1094948	
Northeast	NewEngland	NH	1109718	
Northeast	NewEngland	RI	864989	
Northeast	NewEngland	VT	512375	
sum(states)				100%
West	Pacific	AK	579405	
West	Pacific	CA	31079970	
West	Pacific	HI	1116079	
West	Pacific	OR	3292643	
West	Pacific	WA	2777913	
sum(states)				100%
South	Southatlantic	DC	546229	
South	Southatlantic	DE	752072	
South	Southatlantic	FL	16562355	
South	Southatlantic	GA	8179943	
South	Southatlantic	MD	4503033	
South	Southatlantic	NC	7737873	
South	Southatlantic	SC	3788691	
South	Southatlantic	VA	6347378	
South	Southatlantic	WV	1345706	
sum(states)				100%
Midwest	WNCentral	IA	2377524	
Midwest	WNCentral	KS	2301719	
Midwest	WNCentral	MN	4435973	
Midwest	WNCentral	MO	4914478	
Midwest	WNCentral	ND	600886	
Midwest	WNCentral	NE	1525202	
Midwest	WNCentral	SD	697459	
sum(states)				100%
South	WSCentral	AR	2245902	
South	WSCentral	LA	3472615	
South	WSCentral	OK	2938382	
South	WSCentral	TX	21458360	
sum(states)				100%

A4) Compute a division's share of licensed drivers within its Region (∑(States in a Division)/∑(States in the Corresponding Region)).

Region	Division	Total Division Drivers	Driver % (Division/Region)
Midwest	ENCentral		
Midwest	WNCentral		
Sum(Division)			100%
Northeast	Midatlantic		
Northeast	NewEngland		
Sum(Division)			100%
South	ESCentral		
South	Southatlantic		
South	WSCentral		
Sum(Division)			100%
West	Mountain		
West	Pacific		
Sum(Division)			100%

A5) Compute percentages of licensed drivers by regions among national total, as illustrated below (∑(States in a Region)/∑(States in the Nation)).

Region	Total Regional Drivers	Driver% (Regional/All States)
Midwest		
Northeast		
South		
West		
All		100%

Task A1:
State/∑(All States)

Task A1: SAS Codes (E55a1)

```
LIBNAME mysub 'c:\sas_exercise';
DATA d2 ;
SET mysub.r5d9s50all;
RUN;

PROC SQL;
            CREATE TABLE driverD1 AS
            SELECT *
 ,SUM(Driver) as Total_drivers
 ,Driver/CALCULATED Total_drivers as Driver_Percent_of_nation
FORMAT=percent7.3
             FROM d2;
QUIT;
PROC PRINT DATA =driverD1;
RUN;
```

LIBNAME mysub "c:\sas_exercise" defines a library named mysub representing the folder of c:\sas_exercise.

DATA d2 creates a new SAS dataset named d2.sas7bdat in the default WORK library.

SET mysub.r5d9s50all tells SAS to use data file r5d9s50all.sas7bdata in the mysub library to create the d2 dataset.

RUN kicks off the DATA step.

PROC SQL invokes the SAS SQL function.

*CREATE TABLE driverD1 AS establishes a new data table called driverD1 by selecting all columns (variables) of data (SELECT *) from the dataset d2 created earlier (FROM d2).*

In addition, the new data table driverD1 will have two new variables: (a) Total_drivers is computed using the SUM command to add drivers from all states together (SUM(Driver) as Total_drivers), and (b) Driver_Percent_of_nation is obtained by dividing the number of drivers in a state by national total drivers (Driver/ CALCULATED Total_drivers as Driver_Percent_of_nation).

Pay attention to the word CALCULATED. It is a keyword used by SAS indicating that the Total_drivers is a previously calculated variable and does not need to be recalculated. The use of the CALCULATED keyword must be within the same SELECT statement.

QUIT terminates the PROC SQL procedure.

PROC PRINT DATA=driverD1 tells SAS to print the dataset/table called driverD1 in the WORK library.

RUN kicks off the PROC PRINT procedure.

Task A1 Result Illustration: First 19 rows (states)
State/Σ(All States)

Obs	Region	Division	State	Driver	Registration	Total_drivers	Driver_Percent_of_nation
1	Midwest	ENCentral	IN	5040991	Yes	249258318	2.02%
2	Midwest	ENCentral	IL	9597855	No	249258318	3.85%
3	Midwest	ENCentral	MI	7529922	Yes	249258318	3.02%
4	Midwest	ENCentral	OH	8805656	Yes	249258318	3.53%
5	Midwest	ENCentral	WI	4379360	Yes	249258318	1.76%
6	Midwest	WNCentral	IA	2377524	Yes	249258318	.954%
7	Midwest	WNCentral	KS	2301719	Yes	249258318	.923%
8	Midwest	WNCentral	MN	4435973	No	249258318	1.78%
9	Midwest	WNCentral	MO	4914478	Yes	249258318	1.97%
10	Midwest	WNCentral	ND	600886	Yes	249258318	.241%
11	Midwest	WNCentral	NE	1525202	Yes	249258318	.612%
12	Midwest	WNCentral	SD	697459	Yes	249258318	.280%
13	Northeast	Midatlantic	NJ	7042719	No	249258318	2.83%
14	Northeast	Midatlantic	NY	15449288	No	249258318	6.20%
15	Northeast	Midatlantic	PA	10124749	Yes	249258318	4.06%
16	Northeast	NewEngland	CT	2922797	Yes	249258318	1.17%
17	Northeast	NewEngland	MA	5646648	No	249258318	2.27%
18	Northeast	NewEngland	ME	1094948	Yes	249258318	.439%
19	Northeast	NewEngland	NH	1109718	Yes	249258318	.445%

Task A2:
State/∑(States within the Corresponding Region)

Task A2: SAS Codes (E55a2)

```
LIBNAME mysub 'c:\sas_exercise';
DATA d2 ;
SET mysub.r5d9s50all;
RUN;

PROC SQL;
            CREATE TABLE driverD2 AS
            SELECT *
 ,SUM(Driver) as Total_drivers_by_Region
 ,Driver/CALCULATED Total_drivers_by_Region as Driver_Percent_
in_a_Region  FORMAT=percent7.3
             FROM d2
GROUP BY region;
QUIT;
PROC PRINT DATA=driverD2;
RUN;
```

LIBNAME mysub "c:\sas_exercise" defines a library named mysub representing the folder of c:\sas_exercise.

DATA d2 creates a new SAS dataset named d2.sas7bdat in the default WORK library.

SET mysub.r5d9s50all tells SAS to use the data file r5d9s50all.sas7bdat in the mysub library to create the d2 dataset.

RUN kicks off the DATA step.

PROC SQL invokes SAS SQL function.

CREATE TABLE driverD2 AS establishes a new data table called driverD2.sas7bdat (in SAS data format) by selecting all columns (variables) of data (SELECT) from dataset d2 created earlier (FROM d2).*

In addition, the new data table driverD2 will have two new variables: (a) Total_drivers_by_Region is computed by the SUM command in the form of SUM(Driver) and the GROUP BY region statements to add drivers from states within their corresponding region, and (b) Driver_percent_in_a_ Region is computed by dividing a state's number of Drivers by its corresponding regional total drivers (Total_drivers_by_Region) (Driver/CALCULATED Total_drivers_by_Region).

In this example, pay attention to the GROUP BY variable statement and understand how it works.

Also, pay attention to the word CALCULATED. It is a keyword used by SAS indicating that the Total_drivers_by_Region is a previously calculated variable and does not need to be recalculated. The use of the CALCULATED keyword must be within the same SELECT statement.

QUIT terminates the PROC SQL procedure.

PROC PRINT DATA=driverD2 tells SAS to print the dataset/table called driverD2 in the WORK library.

RUN kicks off the PROC PRINT procedure.

Task A2 Result Illustration: The entire Midwest and some of Northeast states State/∑(States within the Corresponding Region)

Obs	Region	Division	State	Driver	Registration	Total_drivers_by_Region	Driver_Percent_in_a_Region
1	Midwest	ENCentral	MI	7529922	Yes	52207025	14.4%
2	Midwest	ENCentral	WI	4379360	Yes	52207025	8.39%
3	Midwest	ENCentral	IN	5040991	Yes	52207025	9.66%
4	Midwest	WNCentral	NE	1525202	Yes	52207025	2.92%
5	Midwest	WNCentral	IA	2377524	Yes	52207025	4.55%
6	Midwest	WNCentral	KS	2301719	Yes	52207025	4.41%
7	Midwest	WNCentral	SD	697459	Yes	52207025	1.34%
8	Midwest	WNCentral	ND	600886	Yes	52207025	1.15%
9	Midwest	WNCentral	MN	4435973	No	52207025	8.50%
10	Midwest	ENCentral	OH	8805656	Yes	52207025	16.9%
11	Midwest	WNCentral	MO	4914478	Yes	52207025	9.41%
12	Midwest	ENCentral	IL	9597855	No	52207025	18.4%
13	Northeast	Midatlantic	NY	15449288	No	44768231	34.5%
14	Northeast	Midatlantic	NJ	7042719	No	44768231	15.7%
15	Northeast	NewEngland	VT	512375	Yes	44768231	1.14%
16	Northeast	NewEngland	RI	864989	Yes	44768231	1.93%
17	Northeast	Midatlantic	PA	10124749	Yes	44768231	22.6%
18	Northeast	NewEngland	ME	1094948	Yes	44768231	2.45%
19	Northeast	NewEngland	NH	1109718	Yes	44768231	2.48%

Task A3:
State/∑(States within the Corresponding Division)

Task 3: SAS Codes (E55a3)

```
LIBNAME mysub 'c:\sas_exercise';
DATA d2 ;
SET mysub.r5d9s50all;
RUN;
PROC SQL;
            CREATE TABLE driverD3 AS
            SELECT *
 ,SUM(Driver) as Total_Drivers_by_Division
 ,Driver/CALCULATED Total_drivers_by_Division as Driver_Percent_
in_a_division  FORMAT=percent7.3
             FROM d2
               GROUP BY Division;
QUIT;
PROC PRINT DATA=driverD3;
RUN;
```

LIBNAME mysub "c:\sas_exercise" defines a library named mysub representing the folder of c:\sas_exercise.

DATA d2 creates a new SAS dataset named d2.sas7bdat in the default WORK library.

SET mysub.r5d9s50all tells SAS to use data file r5d9s50all.sas7bdat in the mysub library to create the d2 dataset.

RUN kicks off the DATA step.

PROC SQL invokes SAS SQL function.

CREATE TABLE driverD3 AS establishes a new data table called driverD3 by selecting all columns (variables) of data (SELECT) from the dataset d2 created earlier (FROM d2).*

In addition, the new data table driverD3 will have two new variables: Total_Drivers_by_Divison is computed by SUM command in the form of the SUM (Drivers) statement to add drivers from states in a Division as requested by the GROUP BY Division statement.

Driver_Percent_in_a_division is computed by dividing the number of Drivers a state has with the corresponding division total Total_Drivers_by_Division as instructed by the statement below. (Driver/CALCULATED Total_drivers_by_Division).

Pay attention to the word CALCULATED. It is a keyword used by SAS, indicating that the Total_drivers_by_Division is a previously calculated variable and does not need to be recalculated. The use of the CALCULATED keyword must be within the same SELECT statement.

QUIT terminates the PROC SQL procedure.

PROC PRINT DATA=driverD3 tells SAS to print the dataset/table called driverD3 in the WORK library.

RUN kicks off the PROC PRINT procedure.

Task A3 Result Illustration: First 19 states

State/Σ(States within the Corresponding Division)

Obs	Region	Division	State	Driver	Registration	Total_Drivers_by_Division	Driver_Percent_in_a_division
1	Midwest	ENCentral	IL	9597855	No	35353784	27.1%
2	Midwest	ENCentral	OH	8805656	Yes	35353784	24.9%
3	Midwest	ENCentral	MI	7529922	Yes	35353784	21.3%
4	Midwest	ENCentral	IN	5040991	Yes	35353784	14.3%
5	Midwest	ENCentral	WI	4379360	Yes	35353784	12.4%
6	South	ESCentral	AL	3998767	Yes	15060080	26.6%
7	South	ESCentral	TN	5264359	No	15060080	35.0%
8	South	ESCentral	MS	2322325	Yes	15060080	15.4%
9	South	ESCentral	KY	3474629	Yes	15060080	23.1%
10	Northeast	Midatlantic	NY	15449288	No	32616756	47.4%
11	Northeast	Midatlantic	PA	10124749	Yes	32616756	31.0%
12	Northeast	Midatlantic	NJ	7042719	No	32616756	21.6%
13	West	Mountain	NM	1566604	Yes	18498433	8.47%
14	West	Mountain	MT	794211	Yes	18498433	4.29%
15	West	Mountain	NV	2268602	Yes	18498433	12.3%
16	West	Mountain	WY	453927	Yes	18498433	2.45%
17	West	Mountain	CO	4258174	No	18498433	23.0%
18	West	Mountain	UT	2483680	Yes	18498433	13.4%
19	West	Mountain	AZ	5361737	Yes	18498433	29.0%

Task A4:
∑(States in a Division)/∑(States in the Corresponding Region)

Task A4: SAS Codes (E55a4)

```
LIBNAME myEXER           'C:\SAS_EXERCISE';
DATA NEW;
SET myEXER.R5D9S50all;
RUN;

PROC SQL;
CREATE TABLE NEWt AS
SELECT *
FROM NEW;
QUIT;

PROC SQL ;
CREATE TABLE newt2 AS
       SELECT NEWt.Region, Division, sum(Driver) AS SUM_d,
          calculated SUM_d/Subtotal_d AS Percent_r_d
format=percent8.2
   FROM NEWt,
        (SELECT Region, SUM(Driver) AS Subtotal_d FROM NEWT GROUP
BY Region) AS NEWt2b
   WHERE NEWT.Region=NEWt2b.Region
   GROUP BY NEWT.Division
   ORDER BY Region, Division;

QUIT;

PROC SORT DATA=newt2 ;
BY Division;
RUN;
DATA newt2_final;
SET newt2;
BY Division;
IF First.Division=1;
RUN;

PROC SORT DATA=newt2_final OUT =myexer.final_p;
BY Region Division;
RUN;
PROC PRINT DATA=myexer.final_p;
RUN;
```

LIBNAME myEXER "c:\sas_exercise" defines a library named myEXER representing the folder of c:\sas_exercise.

DATA NEW creates a new SAS dataset named NEW.sas7bdat in the default WORK library.

SET myEXER.r5d9s50all tells SAS to use the data file r5d9s50all.sas7bdat in the myEXER library to create the NEW dataset.

RUN kicks off the DATA step.

PROC SQL invokes SAS SQL function.

*CREATE TABLE NEWt AS establishes a new data table called NEWt by selecting all columns (variables) of data (SELECT *) from the dataset NEW created earlier (FROM NEW).*

QUIT terminates this PROC SQL procedure.

PROC SQL invokes SAS SQL function.

CREATE TABLE newt2 AS establishes a new data table called newt2 by (1) selecting variables Region, Division, (2) creating a new variable called SUM_d by using the SUM command in the form of SUM(Driver) per Division instructed by the GROUP BY NEWT.Division statement (The variable Division is from table NEWt), and (3) creating a new variable called Percent_r_d with a format of percent8.2 by using the formula of calculated SUM_d/ Subtotal_d. The variable Subtotal_d will be further defined. Pay attention to the keyword calculated as it signals to SAS that the SUM_d is a pre-calculated variable and does not need to be recomputed.

FROM NEWt is part of the SELECT statement used with the above Selection.

(SELECT Region, sum(Rriver) AS Subtotal_d FROM NEWT GROUP BY Region) AS NEWt2b is a complete set of statements by itself. It creates a new table called NEWt2b by using information from the NEWt table through (1) selecting variable Region and (2) creating a new variable called Subtotal_d by summing drivers up from states in the same Region directed through the GROUP BY Region statement.

WHERE NEWT.Region=NEWt2b.Region merges the two tables NEWT and NEWt2b based on Region.

GROUP BY NEWT.Division is used by the first SELECT statement in creating the table NEWt2.

Order By Region, Division; instruct SAS to list the variable Region first and then Division.

QUIT terminates the PROC SQL procedure.

PROC SORT invokes SAS to sort the newt2.sas7bdat data by Division.

RUN kicks off the PROC SORT procedure.

DATA newt2_final creates a new dataset called newt2_final.sas7bdaat in the WORK library.

SET newt2 instructs SAS to use the newt2.sas7bdat in the WORK folder for the creation.

The BY division used with the above SET statement automatically creates two temporary variables called First.Division and Last.Division (I did not list this one out because I am not using it). Here the condition of IF First.Division=1 asks SAS to keep a row with the file newt2_final only if the row meets the First.Division=1 condition (other than the first row in a given sorted group, others First.Division=0).

PROC SORT DATA=newt2_final sorts the data by region first and then division and delivers the result to the library of myexer with a file name of final_p.sas7bdat.

RUN kicks off the PROC SORT procedure.

PROC PRINT DATA=myexer.final_p prints out the permanent file final_p.sas7bdat in the myexer library.

RUN kicks start the PROC PRINT procedure.

Task A4 Result: All regions
Σ(States in a Division)/Σ(States in the Corresponding Region)

Obs	Region	Division	SUM_d	Percent_r_d
1	Midwest	ENCentral	35353784	67.72%
2	Midwest	WNCentral	16853241	32.28%
3	Northeast	Midatlantic	32616756	72.86%
4	Northeast	NewEngland	12151475	27.14%
5	South	ESCentral	15060080	15.86%
6	South	Southatlantic	49763280	52.42%
7	South	WSCentral	30115259	31.72%
8	West	Mountain	18498433	32.26%
9	West	Pacific	38846010	67.74%

Task A5:
Σ(States in a Region)/Σ(States in the Nation)____

Task A5: SAS Codes (E55a5)

```
LIBNAME myEXER 'C:\
SAS_EXERCISE';
DATA NEW;
SET myEXER.
R5D9S50all;
RUN;

PROC SQL;
CREATE TABLE NEWt AS
SELECT *
FROM NEW;
QUIT;

PROC SQL_;
CREATE TABLE newt3
AS
       SELECT NEWt.
Region, sum(Driver)
AS SUM_By_Region,
          calculated
SUM_By_Region/Total_
all AS Percent_By_
Region
format=percent8.2
```

LIBNAME myEXER "C:\SAS_EXERCISE" defines a library named myEXER representing the folder of C:\SAS_EXERCISE.
DATA NEW creates a new SAS dataset named NEW. sas7bdat in the default WORK library.
SET myEXER.R5D9S50all tells SAS to use data file R5D9S50all.sas7bdat in the myEXER library to create the NEW dataset.
RUN kicks off the DATA step.

PROC SQL invokes SAS SQL function.
*CREATE TABLE NEWt AS establishes a new data table called NEWt by selecting all columns (variables) of data (SELECT *) from the dataset NEW created earlier (FROM NEW).*
QUIT terminates the PROC SQL procedure.
PROC SQL invokes SAS SQL function.
CREATE TABLE newt3 AS establishes a new data table called newt3 by (1) selecting the variable region, (2) creating a new variable called SUM_By_Region by using the command of SUM in the form of sum(Driver) covering a region instructed by the GROUP BY NEWt.Region statement, and (3) creating a new variable called Percent_By_Region with the format of percent8.2 by using the formula of calculated SUM_By_Region/Total_all. Pay attention to the keyword calculated as it tells SAS that the variable Sum_By_Region is pre-calculated and does not need to be recomputed.

```
   FROM NEWt,
        (SELECT
sum(Driver) AS
Total_all FROM
NEWt )
   GROUP BY NEWt.
Region
   ORDER BY
Region;

QUIT;

PROC_SORT
DATA=newt3 ;
BY Region;
RUN;
DATA newt3b;
SET newt3;
BY Region;
IF First.Region=1;
RUN;

PROC SORT
DATA=newt3b OUT
=myexer.final_p3b;
BY Region;
RUN;
PROC PRINT
DATA=myexer.
final_p3b;
RUN;
```

The new variable Total_all is further defined below.

FROM NEWt is part of the SELECT statement used earlier in creating the newt3 table.

(SELECT sum(Driver) AS Total_all FROM NEWt) is a complete statement by itself, and it creates a new variable Total_all FROM NEWt by summarizing Drivers from every State.

GROUP BY NEWt.Region is used by the first SELECT statement in creating the table newt3.

QUIT terminates the PROC SQL procedure.

PROC SORT DATA invokes SAS to sort the newt3.sas7bdat data by Region

RUN kicks off the PROC SORT procedure.

DATA newt3b creates a new dataset called newt3b.sas7bdat in the WORK library.

SET newt3 instructs SAS to use the newt3.sas7bdat in the WORK folder for the creation.

The BY region used with the above SET statement automatically creates two temporary variables called First.region and Last.region (I did not list the Last.Region as there is no need of it in this particular analysis).

Here the condition of IF First.Region=1 asks SAS to keep a row with the file newt3b only if the row meets the condition of First.Region=1.

RUN kicks off the DATA step.

PROC SORT DATA=newt3b sorts the data by Region and delivers it to the library of myexer with a file name of final_p3b.sas7bdat.

RUN kicks off the PROC SORT procedure.

PROC PRINT DATA=myexer.final_p3b prints out the permanent file final_p3b.sas7bdat in the myexer library.

RUN kicks start the PROC PRINT procedure.

Task A5 Result:
$\sum$(States in a Region)/($\sum$(States in the Nation)

Obs	Region	SUM_By_Region	Percent_By_Region
1	Midwest	52207025	20.94%
2	Northeast	44768231	17.96%
3	South	94938619	38.09%
4	West	57344443	23.01%

Task B (dealing with a categorical variable – registration):

B1) Compute percentages of "Yes" and "No" States by Divisions in a Region, as illustrated below:
Σ(Yes States in a Division)/Σ(Yes+No States in the corresponding Division), and
Σ(No states in a Division)/Σ(Yes+No States in the corresponding Division)

Region	Division	Registration	%
Northeast	Midatlantic	No	
Northeast	Midatlantic	Yes	
		SUM (Yes and No) =	**100%**
Northeast	NewEngland	Yes	
Northeast	NewEngland	No	
		SUM (Yes and No) =	**100%**
West	Mountain	Yes	
West	Mountain	No	
		SUM (Yes and No) =	**100%**
West	Pacific	Yes	
West	Pacific	No	
		SUM (Yes and No) =	**100%**
South	Southatlantic	No	
South	Southatlantic	Yes	
		SUM (Yes and No) =	**100%**
South	WSCentral	Yes	
South	WSCentral	No	
		SUM (Yes and No) =	**100%**
Midwest	WNCentral	Yes	
Midwest	WNCentral	No	
		SUM (Yes and No) =	**100%**

B2) Compute % of "No States" by Divisions among all "No States" and "Yes States" by Divisions among all "Yes States" as illustrated below:
Σ(No States in a Division)/Σ(No States with all Divisions), and
Σ(Yes States in a Division)/Σ(Yes States with all Divisions)

Region	Division	Registration	%
Midwest	ENCentral	No	
Midwest	WNCentral	No	
Northeast	Midatlantic	No	
South	ESCentral	No	
South	Southatlantic	No	
South	WSCentral	No	
West	Pacific	No	
West	Mountain	No	
	Sum (Division Nos) =		**100%**
Midwest	ENCentral	Yes	
South	ESCentral	Yes	
Northeast	Midatlantic	Yes	
West	Pacific	Yes	
South	Southatlantic	Yes	
Midwest	WNCentral	Yes	
South	WSCentral	Yes	
West	Mountain	No	
	Sum (Division Yeses) =		**100%**

B3) Compute percentages of "No" States by Region among all "No" States and percentages of "Yes" States by Region among all "Yes" States as illustrated below.
Σ(No States in a Region)/Σ(No States with all Regions), and
Σ(Yes States in a Region)/Σ(Yes States with all Regions)

Region	Registration	%
Midwest	No	
Northeast	No	
South	No	
West	No	
	Sum (Nos) =	**100%**
Midwest	Yes	
South	Yes	
Northeast	Yes	
West	Yes	
	Sum (Yeses) =	**100%**

B4) Compute percentages of "Yes" or "No" States within a given region as illustrated below.
Σ(No States in a Region)/Σ(No + Yes States in the Corresponding Region), and
Σ(Yes States in a Region)/Σ(No + Yes States in the Corresponding Region)

Region	Registration	%
Midwest	No	
Midwest	Yes	
	Sum(Yes and No)=	**100%**
Northeast	No	
Northeast	Yes	
	Sum(Yes and No)=	**100%**
South	No	
South	Yes	
	Sum(Yes and No)=	**100%**
West	No	
West	Yes	
	Sum(Yes and No)=	**100%**

Task B1:
Σ(Yes States in a Division)/Σ(Yes + No states in the corresponding Division)
Σ(No states in a Division/Σ(Yes + No States in the corresponding Division)

Task B1: SAS Codes (E55b1)

```
LIBNAME task4 'C:\SAS_EXERCISE';
DATA NEW;
SET task4.R5D9S50all;
RUN;

PROC SQL;
CREATE TABLE NEWt AS
```

```
SELECT *
FROM NEW;
QUIT;

PROC SQL;
CREATE TABLE task4b AS
   SELECT NEWt.Division, Region, Registration, count(Registration)
AS CountDiv,
          calculated CountDiv/SubtotalD AS PercentYNperRegion
format=percent8.2
   FROM NEWt,
        (SELECT Division, count(*) AS SubtotalD FROM NEWt GROUP BY
Division) AS NEWt2
   WHERE NEWt.Division=NEWt2.Division
   GROUP BY NEWt.Division,Registration
   ORDER BY Region, Registration;
QUIT;

PROC SORT DATA=task4b NODUPKEY OUT=task4.new_sorted_data ;
BY Region Division Registration;
RUN;

PROC PRINT DATA =task4.new_sorted_data;
RUN;
```

LIBNAME task4 "C:\SAS_EXERCISE" defines a library named task4 representing the folder of C:\SAS_EXERCISE.

DATA NEW creates a new SAS dataset named NEW.sas7bdat in the default WORK library.

SET task4.R5D9S50all tells SAS to use data file R5D9S50all.sas7bdat in the mysub library to create the NEW dataset.

RUN kicks off the DATA step.

PROC SQL invokes SAS SQL function.

*CREATE TABLE NEWt AS establishes a new data table called NEWt by selecting all columns (variables) of data (SELECT *) from the dataset NEW created earlier.*

QUIT kicks off the PROC SQL procedure.

PROC SQL invokes SAS SQL function.

CREATE TABLE task4b AS establishes a new data table called task4b by (1) selecting variables Division, Region, and Registration (SELECT NEWt.Divsion, Region, Registration) from the dataset NEWt created earlier, (2) creating a new variable called CountDiv by the function count in the form of count(Registration) with the grouping order of Division first and then Registration (GROUP BY NEWt.Division, Registration. It results 2 types of counts for a Division: number of Yes counts for Registration and number of No counts for the Registration, (3) creating a new variable called PercentYNperRegion in the format of percent8.2 by following the formula of calculated CountDiv/SubtotalD. The SubtotalD is to be defined further.

FROM NEWt instructs all the above SQL steps are carried out with the NEWt dataset.

(SELECT Division, count(*) AS SubtotalD FROM NEWt GROUP BY Division) AS NEWt2 is a complete set of statements by itself. It creates a new table called NEWt2. With this new data table, it has the variable of Division, the count (*) computes the total number of rows (both Yes row and No row) in a Division(the number of States in a Division).

WHERE NEWt.Division=NEWt2.Division merges the NEWt and NEWt2 tables based on Division.

ORDER BY Region, Registration arranges the variable of Region ahead of Registration.

QUIT terminates the PROC SQL procedure.

PROC SORT DATA=task4b NODUPKEY OUT=task4.new_sorted_data creates a new permanent dataset stored in the library of task4 with a file named new_sorted_data. sas7bdat where all duplicate rows are removed from the original task4b dataset. The criteria for duplicate rows are rows that have similar Region, Division, and Registration (BY Region Division Registration).

RUN kicks off the PROC SORT procedure.

PROC PRINT DATA=task4.new_sorted_data instructs SAS to print the new_sorted_data. sas7bdat in the library of task4 on the screen.

RUN kicks off the PROC PRINT procedure.

Task B1 Result:

Σ(Yes States in a Division)/Σ(Yes+No States in the corresponding Division)
Σ(No states in a Division)/Σ(Yes+No States in the corresponding Division)

Obs	Division	Region	Registration	CountDiv	PercentYNperRegion
1	ENCentral	Midwest	No	1	20.00%
2	ENCentral	Midwest	Yes	4	80.00%
3	WNCentral	Midwest	No	1	14.29%
4	WNCentral	Midwest	Yes	6	85.71%
5	Midatlantic	Northeast	No	2	66.67%
6	Midatlantic	Northeast	Yes	1	33.33%
7	NewEngland	Northeast	No	1	16.67%
8	NewEngland	Northeast	Yes	5	83.33%
9	ESCentral	South	No	1	25.00%
10	ESCentral	South	Yes	3	75.00%
11	Southatlantic	South	No	3	33.33%
12	Southatlantic	South	Yes	6	66.67%
13	WSCentral	South	Yes	4	100.0%
14	Mountain	West	No	1	12.50%
15	Mountain	West	Yes	7	87.50%
16	Pacific	West	No	3	60.00%
17	Pacific	West	Yes	2	40.00%

Task B2:

∑(No States in a Division)/∑(No States with all Divisions), and
∑(Yes States in a Division)/∑(Yes States with all Divisions)

Task B2: SAS Codes (E55b2)

```
LIBNAME task4 'C:\SAS_EXERCISE';
DATA NEW;
SET task4.R5D9S50all;
RUN;

PROC SQL;
CREATE TABLE NEWt AS
SELECT *
FROM NEW;
QUIT;

PROC SQL;
CREATE TABLE task4c AS
    SELECT NEWt.Registration,Region,Division, count(Division) AS
Count3,
          calculated Count3/SubtotalD AS Percent format=percent8.2
    FROM NEWt,
        (SELECT Registration, count(*) AS SubtotalD FROM NEWt
        GROUP BY Registration) AS NEWt3
   WHERE NEWt.Registration=NEWt3.Registration
   GROUP BY NEWt.Registration, Rivision;
QUIT;

PROC SORT DATA=task4c NODUPKEY OUT=task4.Task4c_final ;
BY Registration Division Region;
RUN;

PROC PRINT DATA =task4.task4c_final;
RUN;
```

LIBNAME task4 "C:\SAS_EXERCISE" defines a library named task4 representing the folder of C:\SAS_EXERCISE.

DATA NEW creates a new SAS dataset named NEW.sas7bdat in the default WORK library.

SET task4.R5D9S50all tells SAS to use data file R5D9S50all.sas7bdat in the mysub library to create the NEW dataset.

RUN kicks off the DATA step.

PROC SQL invokes SAS SQL function.

CREATE TABLE NEWt AS establishes a new data table called NEWt by selecting all columns (variables) of data (SELECT) from the dataset NEW created earlier.*

QUIT kicks off the PROC SQL procedure.

PROC SQL invokes SAS SQL function.

CREATE TABLE task4c AS establishes a new data table called task4c by
1) Selecting variables Registration, Region, and Division (SELECT NEWt.Registration, Region, Division) from the dataset Newt;
2) Creating a new variable called Count3 by the function Count in the form of count(Division) with the grouping order of Registration first and then Division (GROUP BY NEWt.Registration, Division) leading to separate subtotal Yes counts and subtotal No counts for a Division;
3) Creating a new variable called Percent with the format of percent8.2 by following the formula of calculated Count3/SubtotalD, where the SubtotalD will be defined later. Pay attention to the keyword calculated as it tells SAS that the variable Count3 is pre-calculated and does not need to be recomputed.

FROM NEWt instructs SAS to carry out the above task with the NEWt dataset.

(SELECT Registration, count(*) AS SubtotalD FROM NEWt GROUP BY Registration) AS NEWt3 creates a new table called NEWt3.

With this new NEWt3 table, it has the variable of Registration, the counts of the Registration by Registration types (Total Yes counts and total No counts among all States).

WHERE NEWt.Registration=NEWt3.Registration merges the NEWt and NEWt3 tables based on Registration.

GROUP BY NEWt.Registration, Division is used by the first Select statement.

QUIT terminates the PROC SQL procedure.

PROC SORT Data=task4c NODUPKEY OUT=task4.Task4c_final creates a new permanent dataset stored in the library of Task4c with a file named Task4c_final.sas7bdat where all duplicate rows are removed in the original Task4c dataset. The criteria for duplicate rows are rows having similar Region, Division, and Registration (BY Registration Division Region).

RUN kicks off the PROC SORT procedure.

PROC PRINT DATA=task4.task4c_final instructs SAS to print the task4c_final.sas7bdat in the library of task4 on the screen.

RUN kicks off the PROC PRINT procedure.

Task B2 Result:

Σ(No States in a Division)/Σ(No States with all Divisions), and
Σ(Yes States in a Division)/Σ(Yes States with all Divisions)

Obs	Registration	Region	Division	Count3	Percent
1	No	Midwest	ENCentral	1	7.69%
2	No	South	ESCentral	1	7.69%
3	No	Northeast	Midatlantic	2	15.38%
4	No	West	Mountain	1	7.69%
5	No	Northeast	NewEngland	1	7.69%
6	No	West	Pacific	3	23.08%
7	No	South	Southatlantic	3	23.08%
8	No	Midwest	WNCentral	1	7.69%
9	Yes	Midwest	ENCentral	4	10.53%
10	Yes	South	ESCentral	3	7.89%
11	Yes	Northeast	Midatlantic	1	2.63%
12	Yes	West	Mountain	7	18.42%
13	Yes	Northeast	NewEngland	5	13.16%
14	Yes	West	Pacific	2	5.26%
15	Yes	South	Southatlantic	6	15.79%
16	Yes	Midwest	WNCentral	6	15.79%
17	Yes	South	WSCentral	4	10.53%

Task B3:

Σ(No States in a Region)/Σ(No States with all Regions), and
Σ(Yes States in a Region)/Σ(Yes States with all Regions)

Task B3: SAS Codes (E55b3)

```
LIBNAME MYDATA 'C:\SAS_EXERCISE';
DATA NEW;
SET MYDATA.R5D9S50all;
RUN;

PROC SQL;
CREATE TABLE NEWT AS
SELECT *
FROM NEW;
QUIT;
```

```
PROC SQL;
   SELECT NEWT.Registration, Region, count(Region) AS Count99,
          calculated Count99/Subtotal AS Percent format=percent8.2
   FROM NEWT,
        (SELECT Registration, count(*) AS Subtotal FROM NEWT GROUP
BY Registration) AS NEWT2
   WHERE NEWT.Registration=NEWT2.Registration
   GROUP BY NEWT.Registration, Region
   ORDER BY Registration, Region;
QUIT;
```

LIBNAME MYDATA "C:\SAS_EXERCISE" defines a library named MYDATA representing the folder of C:\SAS_EXERCISE.

DATA NEW creates a new SAS dataset named NEW.sas7bdat in the default WORK library.

SET MYDATA.R5D9S50all tells SAS to use data file R5D9S50all.sas7bdat in the MYDATA library to create the NEW dataset.

RUN kicks off the DATA step.

PROC SQL invokes SAS SQL function.

*CREATE TABLE NEWT AS establishes a new data table called NEWT by selecting all columns (variables) of data (SELECT *) from the dataset NEW (FROM NEW) created earlier.*

QUIT kicks off the PROC SQL procedure.

PROC SQL invokes SAS SQL function.

1) Selecting variables Registration, Region by using the statement of SELECT NEWT. Registration, Region from the dataset NEWT;

2) Creating a new variable called Count99 by the Count function in the form of count(Region) with the grouping order of Registration first and then Region (GROUP BY NEWT.Registration, Region) leading to separate subtotal Yes counts and subtotal No counts for each region;

3) Creating a new variable called Percent with the format of percent8.2 by following the formula of calculated Count99/Subtotal, where the Subtotal will be defined later. Pay attention to the keyword calculated as it tells SAS that count99 is pre-computed and does not need to be recalculated.

FROM NEWT instructs SAS to carry out the above task with the NEWT dataset.

(SELECT Registration, count(*) AS Subtotal FROM NEWT GROUP BY Registration) as NEWT2 is a completely self-contained statement and it creates a new table called NEWT2 by selecting the variable of Registration, and creating a new variable called Subtotal through the count command in the form of count(*) of the number of Registrations by Registration types (Yes counts and No counts among all states).

WHERE NEWT.Registration=NEWT2.Registration merges the NEWT and NEWT2 tables based on Registration.

GROUP BY NEWT.Registration, Region is used by the first SELECTstatement.

QUIT terminates the PROC SQL procedure.

Task B3 Result:

Σ(No States in a Region)/Σ(No States with all Regions), and
Σ(Yes States in a Region)/Σ(Yes States with all Regions)

Registration	Region	Count	Percent
No	Midwest	2	15.38%
No	Northeast	3	23.08%
No	South	4	30.77%
No	West	4	30.77%
Yes	Midwest	10	26.32%
Yes	Northeast	6	15.79%
Yes	South	13	34.21%
Yes	West	9	23.68%

Task B4:

Σ(No States in a Region)/Σ(No+Yes States in the Corresponding Region), and
Σ(Yes States in a Region)/Σ(No+Yes States in the Corresponding Region)

Task B4: SAS Codes (E55b4)

```
LIBNAME MYDATA 'C:\SAS_EXERCISE';
DATA NEW;
SET MYDATA.R5D9S50all;
RUN;

PROC SQL;
CREATE TABLE NEWT AS
SELECT *
FROM NEW;
QUIT;

PROC SQL;
   SELECT NEWT.Region, Registration, count(Registration) AS
Count3,
          calculated Count3/Subtotal3 AS Percent format=percent8.2
   FROM NEWT,
         (SELECT Region, count(*) AS Subtotal3 FROM NEWT GROUP BY
Region) AS NEWT3
   WHERE NEWT.Region=NEWT3.Region
   GROUP BY NEWT.Region, Registration
   ORDER BY Region, Registration;
QUIT;
```

LIBNAME MYDATA "C:\SAS_EXERCISE" defines a library named MYDATA representing the folder of C:\SAS_EXERCISE.

DATA NEW creates a new SAS dataset named NEW.sas7bdat in the default WORK library.

SET MYDATA.R5D9S50all tells SAS to use data file R5D9S50all.sas7bdat in the MYDATA library to create the NEW dataset.

RUN kicks off the DATA step.

PROC SQL invokes SAS SQL function.

*CREATE TABLE NEWT AS establishes a new data table called NEWT by selecting all columns (variables) of data (SELECT *) from the dataset NEW (FROM NEW) created earlier.*

QUIT kicks off the PROC SQL procedure.

PROC SQL invokes SAS SQL function.

1) Selecting variables Region, Registration (SELECT NEWT.Region, Registration...) FROM NEWT;

2) Creating a new variable called Count3 by using the Count function in the form of count(Registration) with the grouping order of Region first and then Registration (GROUP BY NEWT.Region, Registration) leading to separate subtotal Yes counts and subtotal No counts for each Region;

3) Creating a new variable called Percent with the format of percent8.2 by following the formula of calculated Count3/Subtotal3, where the Subtotal3 will be defined later. The keyword calculated tells SAS that Count3 is pre-calculated and does not need to be recomputed.

FROM NEWT instructs SAS to carry out the above task with the NEWT dataset.

(SELECT Region, count(*) AS Subtotal3 FROM NEWT GROUP BY Region) AS NEWT3 is an entirely self-contained statement, and it creates a new table called NEWT3 by selecting the Region variable from the NEWT dataset and creating a new variable Subtotal3 through the Count command in the form of count(*), which equals to total numbers of states in a Region.

WHERE NEWT.Region=NEWT3.Region merges the NEWT and NEWT3 tables based on Region.

GROUP BY NEWT.Region, Registration is used by the first SELECT statement.

ORDER BY Region, Registration arranges Region ahead of Registration in the output dataset.

QUIT terminates the PROC SQL procedure.

Task B4 Result:

Σ(No States in a Region)/Σ(No+Yes States in the Corresponding Region), and
Σ(Yes States in a Region)/Σ(No+Yes States in the Corresponding Region)

Region	Registration	Count3	Percent
Midwest	No	2	16.67%
Midwest	Yes	10	83.33%
Northeast	No	3	33.33%
Northeast	Yes	6	66.67%
South	No	4	23.53%
South	Yes	13	76.47%
West	No	4	30.77%
West	Yes	9	69.23%

5.6 COMPUTING MOVING AVERAGE BY USING LAG AND LAGn

Time series data often involve lagged data. While there are many ways to handle lagged data in SAS, the LAG and LAGn commands are the most straightforward. LAGn will return the value of the nth previous time period value.

To compute a moving average, the most important factor that you need to know is the number of periods used in the calculation. For example, a 12-month moving average will need 12 consecutive months of data, and a 3-week moving average will need 3 consecutive weeks of observations. The LAG command takes the general form below.

LAG<n>(variable)

where n is the number of lag periods.

You can use the LAG and LAGn statements to facilitate your moving average computation, as illustrated below.

Example: 5.6

Data: average.sas7bdat

The original data is listed in the "Original" column. Other shaded columns are listed to illustrate the concept of lag statements.

Observation #	Original	Lag(original)	Lag2(original)	Moving Average of 3 Obs
1	6			
2	5	6		
3	4	5	6	
4	3	4	5	4.5
5	1	3	4	3.5
6	10	1	3	2
7	12	10	1	5.5
8	14	12	10	11
9	16	14	12	13

Task: Compute a three-period moving average.

SAS Codes (E56)

```
LIBNAME my_ave 'c:\
sas_exercise';
DATA average2;
SET my_ave.average;
MyFirstLag=LAG(Original);
*Lag1 is the same as Lag;
MySecondLag=LAG2(Original);
My_3_MA= MEAN(Original,
MyFirstLag, MySecondLag);
```

LIBNAME my_ave "c:\sas_exercise" defines a library named my_ave representing the folder of C:\sas_exercise.

DATA average2 creates a SAS dataset named average2.sas7bdat and stores it in the WORK library.

SET my_ave.average instructs SAS to use a data file named average.sas7bdat in the library of my_ave to create the average2 dataset.

```
Alt_My_3_MA=mean(Original,
lag(Original),lag2(Original));

RUN;
PROC PRINT DATA=average2;
RUN;

DATA average_w3 ;
SET average2 (FIRSTOBS=3) ;
PROC PRINT DATA=average_w3;
RUN;
```

MyFirstLag=LAG(Original) creates a new variable called MyFirstLagby the SAS LAG function. LAG(Original) lags 1 period.

MySecondLag=LAG2(Original) creates a new variable called MySecondLag by the SAS LAG function. LAG2(Original) lags 2 period.

My_3_MA=MEAN(Original, MyFirstLag, MySecondLag) creates a new variable named My_3_MA by SAS function mean to compute the average of the three variables.

Alt_My_3_MA=mean(original, lag(Original),lag2(Original)) creates a new variable named Alt_My_3_MA by SAS function mean to compute the average of the original observation, one lag observation, and 2 lag observations through a nested function arrangement vs. creating the new variable approach under the My_3_MA case.

RUN kicks off the DATA step.

PROC PRINT DATA=average2 instructs SAS to print the average2.sas7bdat data in the WORK library on the screen.

RUN kicks off the PROC PRINT step.

DATA average_w3 creates a new dataset named average_w3.sas7bdat in the WORK library.

SET average2 (FIRSTOBS=3) instructs SAS to use the dataset named average2.sas7bdat in the WORK library, starting with row 3 to create the average_w3.sas7bdat file.

PROC PRINT Data=average_w3 tells SAS to print the average_w3.sas7bdat in the WORK library on the screen.

RUN kicks off the PROC PRINT step.

Results:

Obs	Original	MyFirstLAG	MySecondLAG	My_3_MA	Alt_My_3_MA
1	6	.	.	6.0000	6.0000
2	5	6	.	5.5000	5.5000
3	4	5	6	5.0000	5.0000
4	3	4	5	4.0000	4.0000
5	1	3	4	2.6667	2.6667
6	10	1	3	4.6667	4.6667
7	12	10	1	7.6667	7.6667
8	14	12	10	12.0000	12.0000
9	16	14	12	14.0000	14.0000

Obs	Original	MyFirstLAG	MySecondLAG	My_3_MA	Alt_My_3_MA
1	4	5	6	5.0000	5.0000
2	3	4	5	4.0000	4.0000
3	1	3	4	2.6667	2.6667
4	10	1	3	4.6667	4.6667
5	12	10	1	7.6667	7.6667
6	14	12	10	12.0000	12.0000
7	16	14	12	14.0000	14.0000

5.7 PRODUCING TOTAL FOR EACH BY-GROUP VARIABLE

Generating a total for a variable (column) disaggregated by the BY-group is a task you will encounter often. Pay attention to how totals are computed with multidimensional variables and how the order of variables in the **PROC SORT** and **SET By-Group** are affecting the results.

Example: 5.7

Data: bus.sas7bdat
The bus.sas7bdat data has four variables as illustrated below:

- CountyID
- Year
- DW (Day of Week): 1 is Sunday, 2 is Monday... and 7 is Saturday)
- BusMiles (total million miles traveled by transit buses for all days of the week in a year in that County).

CountyID	Year	DW	BusMiles
MT661	14	1	1.98
MT661	14	2	1.75
MT661	14	3	1.36
MT661	14	4	1.66
MT661	14	5	1.5
MT661	14	6	1.69
MT661	14	7	1.99
PG256	14	1	2.08
PG256	14	2	1.84
PG256	14	3	1.43
PG256	14	4	1.75
PG256	14	5	1.58
PG256	14	6	1.78
PG256	14	7	2.09
MT661	15	1	1.57
MT661	15	2	2.57
MT661	15	3	2.37

Tasks:

1. Compute total bus miles by County (regardless of DW and Year).
2. Compute total bus miles by County and DW (regardless of Year).
3. Compute total bus miles by year and DW (regardless of County).

Task1 SAS Codes (E57t1) - Compute total bus miles by County (E57t1)

```
LIBNAME mytotal 'c:\sas_exercise';
PROC SORT DATA=mytotal.BUS;
BY CountyID;
Run;
DATA task1;
   SET mytotal.bus;
   BY countyID;
   IF FIRST.CountyID THEN
totalBus=0;
   totalBus + busMiles;
   IF LAST.CountyID THEN OUTPUT;
   KEEP countyID totalBus;
RUN;
PROC PRINT DATA=task1;
RUN;
```

LIBNAME mytotal "c:\sas_exercise" defines a library named mytotal representing the folder of C:\sas_exercise.

PROC SORT DATA=mytotal.BUS; BY CountyID statements instruct SAS to sort the data file named BUS.sas7bdat in the mytotal library in ascending order (default) by the variable of CountyID.

Run kicks off the PROC SORT step.

DATA task1 creates a SAS dataset named task1.sas7bdat and stores it in the WORK library.

SET mytotal.bus instructs SAS to use a data file named BUS.sas7bdat in the library of mytotal to create the task1 dataset.

When the BY countyID is used with the SET statement, it automatically generates two temporary variables FIRST.CountyID and LAST.CountyID.

The IF FIRST.CountyID THEN totalBus=0 and

totalBus+BusMiles statements kick the start of the creation of a new variable named totalBus and a cumulative counter totalBus+busMiles function.

For example, SAS reads the first data row in the sorted dataset where FIRST.CountyID=1, the totalBus is set to 0, then the totalBus is obtained by adding 0+busMiles (e.g., first row busMiles=3.4) together, which is 3.4.

Now, SAS reads the second row where FIRST.CountyID=0, busMiles=12.3, totalBus now equals to 3.4+12.3 =15.7. This will continue until the LAST.CountyID=1 row is read. The above step will start again with the second CountyID per the sorted records.

KEEP countyID totalBus instructs SAS to only keep the variables of countyID and totalBus in the task1.sas7bdat.

RUN kicks off the DATA step.

PROC PRINT DATA=task1 instructs SAS to print the file task1.sas7bdat in the WORK library on the screen.

RUN kicks off the PROC PRINT step.

Task 1 Result: Bus miles by County

Obs	CountyID	totalBus
1	MT661	85.50
2	PG256	89.92

Task 2 SAS Codes (E57t2) - Compute total bus miles by County and DW

```
LIBNAME mytotal 'c:\sas_exercise';
PROC SORT DATA=mytotal.BUS;
BY CountyID DW;
Run;

DATA task2;
 SET mytotal.BUS;
   BY countyID DW;
   IF FIRST.DW THEN totalBus2=0;
   totalBus2 + busMiles;
   IF LAST.DW THEN OUTPUT;
   KEEP countyID DW totalBus2;
RUN;
PROC PRINT DATA=task2;
RUN;
```

LiBNAME mytotal "c:\sas_exercise" defines a library named mytotal representing the folder of C:\sas_exercise.

PROC SORT DATA=mytotal.BUS; BY CountyID DW statements instruct SAS to sort the data file named BUS.sas7bdat in the mytotal library by the variable of CountyID first, and then by DW, in ascending (default) order.

Run kicks off the PROC SORT step.

DATA task2 creates a SAS dataset named task2.sas7bdat and stores it in the WORK library.

SET mytotal.BUS instructs SAS to use a data file named BUS.sas7bdat in the library of mytotal to create the task2 dataset.

When the BY countyID DW is used with the SET statement, it automatically generates two temporary variables FIRST.DW and LAST.DW.

The IF FIRST.DW THEN totalBus2=0 and

totalBus2+busMiles statements kick the start of the creation of a new variable named totalBus2 and a cumulative counter totalBus2+busMiles function.

When SAS reads the first row in the sorted by countyID and DW file, FIRST.DW=1, the totalBus2 is set to 0, busMiles=3.4 is read (as an example for illustration).

The cumulative counter totalBus2+busMiles calculates the totalBus2 by adding 0 and 3.4 together. The totalBus2=3.4.

Task 2 Result: Bus mile by County and DW

Obs	CountyID	DW	totalBus2
1	MT661	1	10.79
2	MT661	2	11.71
3	MT661	3	11.72
4	MT661	4	12.89
5	MT661	5	13.09
6	MT661	6	13.02
7	MT661	7	12.28
8	PG256	1	11.35
9	PG256	2	12.31
10	PG256	3	12.32
11	PG256	4	13.56
12	PG256	5	13.77
13	PG256	6	13.69
14	PG256	7	12.92

Now, SAS reads the second row where FIRST.DW=0, BusMiles=12.3, totalBus2 = 3.4+12.3 =15.7. This will continue until the LAST.DW=1 condition is met (This is the last row of similar rows per the sort By variables). From here, the above step will start again with the second DW type.

KEEP countyID DW totalBus2 instructs SAS to only keep variables of countyID, DW, and totalBus2 in task2.sas7bdat file.

RUN kicks off the DATA step.

PROC PRINT DATA=task2 instructs SAS to print the file task2.sas7bdat in the WORK library on the screen.

RUN kicks off the PROC PRINT step.

Task 3 SAS Codes (E57t3) - Compute total bus miles by year and DW

```
LIBNAME mytotal 'c:\sas_exercise';
PROC SORT DATA=mytotal.BUS;
BY Year DW;
Run;

DATA task3;
   SET mytotal.bus;
   BY Year DW;
   IF FIRST.DW THEN totalBus3=0;
   totalBus3 + busMiles;
   IF LAST.DW THEN OUTPUT;
   KEEP Year DW totalBus3;
RUN;
PROC PRINT DATA=task3;
RUN;
```

LIBNAME mytotal "c:\sas_exercise" defines a library named mytotal representing the folder of C:\sas_exercise.

PROC SORT DATA=mytotal.BUS; BY Year DW statements instruct SAS to sort the data file named BUS.sas7bdat in the mytotal library by the variable of Year first and then DW in an ascending (default) order.

Run kicks off the PROC SORT step.

DATA task3 creates a SAS dataset named task3.sas7bdat and stores it in the WORK library.

SET mytotal.BUS instructs SAS to use a data file named BUS.sas7bdat in the mytotal library to create the task3 dataset.

When the BY Year DW is used with the SET statement, it automatically generates two temporary variables FIRST.DW and LAST.DW.

The IF FIRST.DW THEN totalBus3=0 and

totalBus3+BusMiles statement kicks the start of the creation of a new variable named totalBus3 and a cumulative counter equation of totalBus3+busMiles.

For example, when SAS reads the first row in the sorted by Year and DW file, FIRST.DW=1, totalBus3 is set to 0, and busMiles=5.4. The counter totalBus3+busMiles then calculates the totalBus3 by adding 0 with 5.4 together. totalBus3=5.4.

Now, SAS reads the second row of the sorted data file where FIRST.DW=0, busMiles=3.1, totalBus3=5.4+3.1 =8.5. This will continue until the LAST.DW=1 row condition (FIRST.DW still equals to 0) is met. After that, the entire step listed above will start again with the second DW type.

KEEP Year DW totalBus3 instructs SAS to only keep variables of Year, DW, and totalBus3.

RUN kicks off the DATA step.

PROC PRINT DATA=task3 instructs SAS to print the file task3.sas7bdat in the WORK library on the screen.

RUN kicks off the PROC PRINT step.

Task 3 Result: Bus miles by year and DW

Obs	Year	DW	totalBus3
1	14	1	4.06
2	14	2	3.59
3	14	3	2.79
4	14	4	3.41
5	14	5	3.08
6	14	6	3.47
7	14	7	4.08
8	15	1	3.22
9	15	2	5.27
10	15	3	4.86
11	15	4	5.91
12	15	5	5.37
13	15	6	5.44
14	15	7	4.31
15	16	1	3.90
16	16	2	4.14
17	16	3	4.29
18	16	4	4.74
19	16	5	4.64

5.8 RETAINING PREVIOUS VALUES AND SUM WITH ORIGINAL DATA

Retaining a previously iterated value during SAS's DATA step with the original data is common during data processing. To retain a value, you can use the RETAIN statement. In addition, to compute and keep a cumulative value (e.g., var1), you can use the expression of "SUM+var1."

Example: 5.8

Data: retain_vmt.sas7bdat: 4 variables

- CountyID
- Year
- Month
- Truck (miles traveled by all trucks in a county during a given month).

A portion of the data is listed below for illustration.

CountyID	Year	Month	Truck
PG256	19	January	8.9
PG256	19	February	10.1
PG256	19	March	8.7
PG256	19	April	10.4
PG256	19	May	12.5
PG256	19	June	9.7
PG256	19	July	11.3
PG256	19	August	9.4
PG256	19	September	9.1
PG256	19	October	11.1
PG256	19	November	11.6
PG256	19	December	11
PG256	18	January	8.8
PG256	18	February	11
PG256	18	March	10.3

Task:

- Compute cumulative truck miles traveled by month and year.
- Identify the maximum monthly truck miles traveled among current and all previous months in a calendar year, as illustrated below. The shaded columns are to be computed by the SAS code file.

CountyID	Year	Month	Truck	Annual Max TrucK	Annual Cumulative Truck
PG256	19	January	8.9	8.9	8.9
PG256	19	February	10.1	10.1	19.0
PG256	19	March	8.7	10.1	27.7
PG256	19	April	10.4	10.4	38.1
PG256	19	May	12.5	12.5	50.6
PG256	19	June	9.7	12.5	60.3
PG256	19	July	11.3	12.5	71.6
PG256	19	August	9.4	12.5	81.0
PG256	19	September	9.1	12.5	90.1
PG256	19	October	11.1	12.5	101.2
PG256	19	November	11.6	12.5	112.8
PG256	19	December	11	12.5	123.8
PG256	18	January	8.8	8.8	8.8
PG256	18	February	11	11	19.8
PG256	18	March	10.3	11	30.1
PG256	18	April	9.5	11	39.6

SAS Codes (E58)

```
LIBNAME myretain 'c:\sas_exercise';
data vmt ;
set myretain.retain_vmt ;
RUN;
```

```
PROC SORT DATA=vmt;
BY countyID year month;
RUN;

DATA r2;
SET vmt;
BY countyID year month;
IF FIRST.year THEN
DO;
Cumtruck=0;
Truckmax=truck;
END;
RETAIN truckmax;
truckmax= MAX(truckmax, truck);
Cumtruck+truck;
RUN;

PROC PRINT DATA= r2;
RUN;
```

LIBNAME myretain "c:\sas_exercise" defines a library named myretain representing the folder of c:\sas_exercise.

Data vmt creates a new dataset called vmt.sas7bdat to be stored in the default WORK library.

set myretain.retain_vmt instructs SAS to use the data file named vmt.sas7bdat in the myretain library to create the vmt file.

RUN kicks off the DATA step.

PROC SORT DATA=vmt; BY countyID year month statements instruct SAS to sort the data file named vmt in the WORK library by the order of countyID, year, and month in an ascending order.

RUN kicks off the PROC SORT procedure.

DATA r2 creates a new data file named r2 in the WORK folder.

SET vmt instructs SAS to use the vmt.sas7bdat file in the WORK library to create the r2 dataset.

BY countyID Year month used with the above SET statement together automatically triggers the generation of 2 temporary variables FIRST.year and LAST.year (the LAST.year is not used in the example here).

IF FIRST.year THEN is a conditional statement stating that in the sorted data file, if the FIRST.year=1, proceed with the instruction below.

DO; Cumtruck=0; Truckmax=truck; END; is a DO loop statement. It instructs SAS if the FIRST.year=1, then creates the Cumtruck variable and set it to 0, creates the Truckmax variable and set it equals to truck. The rationale is that the FIRST.year=1 observation row is the first row for that unique variable based on the sorted result.

RETAIN truckmax tells SAS to keep the truckmax variable in the dataset and update the variable truckmax by using the SAS function of MAX (truckmax, truck) for values between truckmax and current truck. Truckmax takes the larger value of these two numbers.

Cumtruck+truck is an accumulator function where cumulative truck miles is calculated per sorted group. For example, when FIRST.year=1 row is read, Cumtruck=0, truck=2.5, the accumulator function calculates the Cumtruck=0+2.5=2.5.

When SAS reads the second row where FIRST.year=0, truck =6.1, the Cumtruck =2.5+6.1=8.6.

The above steps keep on going until a new FIRST.year=1 condition is met, indicating that a new BY Group (different year) row is encountered. Now, the Cumtruck is rest to 0, and truck is read again. More rows are read and corresponding calculations are made.

The process continues until all BY groups are read.

RUN kicks off the DATA step.

PROC PRINT DATA=r2 instructs SAS to print out the r2.sas7bdat in the WORK library.

RUN kicks off the PROC PRINT procedure.

Result Illustration: Cumulative VMT by month and year and maximum monthly VMT for current and previous months within a calendar year.

Obs	CountyID	Year	Month	Truck	Cumtruck	Truckmax
1	BA887	17	1	12.2	12.2	12.2
2	BA887	17	2	14.0	26.2	14.0
3	BA887	17	3	9.8	36.0	14.0
4	BA887	17	4	10.0	46.0	14.0
5	BA887	17	5	10.6	56.6	14.0
6	BA887	17	6	11.0	67.6	14.0
7	BA887	17	7	12.1	79.7	14.0
8	BA887	17	8	11.8	91.5	14.0
9	BA887	17	9	13.5	105.0	14.0
10	BA887	17	10	11.4	116.4	14.0
11	BA887	17	11	8.2	124.6	14.0
12	BA887	17	12	13.5	138.1	14.0
13	BA887	18	1	11.0	11.0	11.0
14	BA887	18	2	9.3	20.3	11.0
15	BA887	18	3	9.0	29.3	11.0
16	BA887	18	4	10.8	40.1	11.0
17	BA887	18	5	11.3	51.4	11.3
18	BA887	18	6	10.7	62.1	11.3
19	BA887	18	7	8.6	70.7	11.3

5.9 COLLAPSING OBSERVATIONS WITH NEW OUTPUT DATA

Data dimension reductions (eliminating certain variables) are common occurrences in both data analysis and data sharing.

For example, you have a dataset containing confidential individual household income information by neighborhood in a city. You don't want to give such data out due to privacy concerns. By collapsing individual household observations into their corresponding neighborhoods, you can release the neighborhood data without the privacy issue concerning individual households. During data dimension reduction, pay attention to the total sum, which should not be changed.

Example: 5.9

Data: collpsping.sas7bdat
The data has four variables: County ID, year, month, and truck miles, as illustrated below.

CountyID	Year	Month	Truck
PG256	19	January	8.9
PG256	19	February	10.1
PG256	19	March	8.7
PG256	19	April	10.4
PG256	19	May	12.5
PG256	19	June	9.7
PG256	19	July	11.3
PG256	19	August	9.4
PG256	19	September	9.1
PG256	19	October	11.1
PG256	19	November	11.6
PG256	19	December	11
PG256	18	January	8.8
PG256	18	February	11
PG256	18	March	10.3

Task: Create a new dataset without the year variable by aggregating truck miles from all years, as illustrated below.

CountyID	Month	CumTruck
PG256	January	32.9
PG256	February	30.8
PG256	March	28
PG256	April	29.1
PG256	May	31.3
PG256	June	32.7
PG256	July	29.4
PG256	August	32.5
PG256	September	32.1
PG256	October	31
PG256	November	32
PG256	December	33.2
PG256	January	32
PG256	February	29.9
PG256	March	27.1
PG256	April	28.3

SAS Codes (E59)

```
LIBNAME myretain 'c:\sas_exercise';
data vmt;
set myretain.collapsing ;
RUN;

PROC SORT DATA=vmt;
BY countyID month;
RUN;
```

LIBNAME myretain "c:\sas_exercise" defines a library named myretain representing the folder of c:\sas_exercise.

DATA vmt creates a new dataset called vmt.sas7bdat in the WORK library.

set myretain.collapsing instructs SAS to use the data file named collapsing.sas7bdat in the myretain library to create the vmt file.

RUN kicks off the DATA step.

PROC SORT DATA=vmt; by countyID month statements instruct SAS to SORT the data file named vmt.sas7bdat in the WORK library by the order of countyID first and then month in an ascending order.

RUN kicks off the PROC SORT procedure.

DATA collapse2 creates a new data file named collapse2 in the WORK library.

SET vmt instructs SAS to use the vmt.sas7bdat file in the WORK library to create the collapse2 dataset.

```
DATA collapse2;
SET vmt;
BY countyID month;
IF FIRST.month THEN
DO;
Cumtruck=0;
END;
Cumtruck+truck;
IF LAST.month THEN
DO;
OUTPUT;
END;
KEEP countyID month CumTruck
RUN;

PROC sort data=collapse2;
by countyid month;
run;

PROC PRINT DATA=collapse2;
RUN;
```

Result:

Obs	CountyID	Month	Cumtruck
1	BA887	1	32.9
2	BA887	2	30.8
3	BA887	3	28.0
4	BA887	4	29.1
5	BA887	5	31.3
6	BA887	6	32.7
7	BA887	7	29.4
8	BA887	8	32.5
9	BA887	9	32.1
10	BA887	10	31.0
11	BA887	11	32.0
12	BA887	12	33.2
13	MT661	1	32.0
14	MT661	2	29.9
15	MT661	3	27.1
16	MT661	4	28.3
17	MT661	5	30.4
18	MT661	6	31.8
19	MT661	7	28.7

BY countyID month used with the above SET statement together automatically triggers the generation of 2 temporary variables FIRST.month and LAST.month

IF FIRST.month THEN is a conditional statement stating that in the sorted dataset, if the FIRST.month=1, then proceed with the instruction below.

DO; Cumtruck=0; END; Cumtruck+truck; statements instruct SAS if the FIRST.month=1, then creates the Cumtruck variable and set it to 0. Use the accumulator statement Cumtruck+truck to calculate Cumtruck. The rational is that the FIRST.month=1 observation row is the first row for that unique observation row based on the sorted result.

For example, when FIRST.month=1 row is read, Cumtruck=0, truck=1.2, the accumulator function calculates the Cumtruck=0+1.2=1.2.

When SAS reads the second row where FIRST.month=0, truck =5.7, the Cumtruck =1.2+5.7=6.9.

END terminates the DO loop.

The above step keeps going until the IF LAST.month then (similar to If LAST.month=1 then) condition is met.

Once the "IF LAST.Month THEN" condition is met, SAS is instructed to output the file until all the BY groups are read.

END terminates the DO loop.

KEEP countyID month CumTruck instructs SAS to keep variables countyID month CumTruck in the collapse2.sas7bdat dataset.

RUN kicks off the DATA step.

PROC PRINT DATA=collapse2 instructs SAS to print out the collapse2.sas7bdat in the WORK library on the screen.

RUN instructs SAS to execute the PROC PRINT procedure.

6

Common Statistical Procedures

SAS's statistical analysis procedures are both powerful and easy to use. Performing actual data analysis through SAS is likely to be your least challenging issue. The challenges you are most likely to encounter in the statistical analysis include:

- Designing your experiment
- Collecting data
- Selecting an appropriate statistical procedure
- Interpreting results correctly.

This chapter covers basic statistical concepts and their associated statistical SAS procedures.

6.1 p VALUE AND ALPHA LEVEL

6.1.1 p Value

To determine whether there is a difference (statistical significance) among means of different populations represented by your data (sample), or whether your regression coefficients or correlation coefficients equal 0, you rely on a number called the p value.

So, what is a p value? A p value is a probability of obtaining a value of the test statistic that is as extreme as, or more extreme than, the value given that the null hypothesis is true. To fully understand the sentence, we must first understand the basics of hypothesis testing.

Hypothesis testing is used in statistical analysis to determine the probability that a hypothesis is true. The two major components of hypothesis testing are the null hypothesis and the alternative hypothesis. The null hypothesis, notated as Ho, is the hypothesis that there is no statistically significant difference between two populations (e.g., means of two populations) and that any difference that does exist is by random chance, or through sampling or experimental error. The alternative hypothesis, denoted as H1 or Ha, is the hypothesis that differences in observations are not the result of random chance.

All statistical significance tests have this equality null hypothesis. For example, with the t-test and F-test, the null hypothesis is that all means are equal. For regression analysis, the null hypothesis is that coefficients for your independent variables (e.g., x_1, x_2 ...) equal zero. And for your correlation analysis, coefficients of correlation equal zero.

Now, let's examine the concepts of differences and variations. Differences and variations can be either "real" or "random chance." The p value enables us to judge how real a difference is between "real" and "random chance." If the probability for an observed difference due to "random chance" is high (e.g., 90%), it is unlikely there is a "real" difference. On the other hand, if the "random chance" probability is low (e.g., 1%), then it is telling us that the difference observed is more likely "real."

Now, let's use a p value = 0.9876 from a t-test of population A and B $\left(\bar{A}=10.7 \text{ and } \bar{B}=10.9\right)$. This p value indicates that the probability of seeing the mean difference between Population A and B $\left(\bar{A}-\bar{B}=10.7-10.9=-0.2\right)$ or larger due to "random chance" is 98.76%. This is very strong evidence that the difference we see is mainly due to "random chance." We fail to reject the null hypothesis, which is that there is no statistically significant difference between A and B.

Now, let's say the p value is 0.0123 for the above t-test. This new p value tells us that the probability of seeing the mean difference between Population A and Population B to be −0.2 or larger due to "random chance" is 1.23%. This is very strong evidence that the difference we see is highly unlikely from "random chance." We should reject the null hypothesis because we know there is only a 1.23% chance that the difference is random.

6.1.2 p Value Interpretation

Typical acceptable levels of p value used in rejecting the null hypothesis are listed below:

- p value = <0.001, there is highly significant and overwhelming evidence against the null hypothesis.
- 0.001 < p value = <0.01, there is strong and significant evidence against the null hypothesis.
- 0.01 < p value = <0.05, there is significant moderate evidence against the null hypothesis.
- 0.05 < p value = <0.1, there is some significant evidence against the null hypothesis.
- 0.1 < p value, there is no significant evidence against the null hypothesis (failed to reject hypothesis).

There are no precise delineations for the p value. The above-listed value ranges are nothing but some practical guidelines. The best strategy is to always present your p value vs. the usage of a predefined α-level.

6.1.3 Alpha Level

The alpha level (α-level) is called the significance level. It is the probability level to reject the null hypothesis when your p value is less than or equal to it.

The establishment of an α-level is an arbitrary decision by researchers themselves. When the p value is larger than a predefined α-level, you fail to reject your hypothesis. When your p value is smaller than or equal to the α-level, you reject your hypothesis.

The language used is illustrated below:

Case 1 Scenario Illustration: Let's say your p value is 0.0312. And your α-level = 5%:

At a 5% significance level, there is a significant difference between M and N with N being 8.4 higher than M.

Case 2 Scenario Illustration: Let's say your p value is 0.1234. And your α-level = 5%:

At a 5% significance level, there is no significant difference between M and N.

6.2 DATA DISTRIBUTION AND DESCRIPTIVE STATISTICS – PROC UNIVARIATE

Many of the most common methods for statistical analysis (t-test, Z-test, F-test, GLM (general linear model), etc.) rely on the assumption that the underlying data are normally distributed. If the data are not normally distributed, there are methods you can use to attempt to transform the data into a normal distribution.

The desire for normal distribution rests with the availability of the aforementioned precise and highly reliable statistical methods. These classical statistical procedures can calculate parameters such as mean, variance, standard deviation, and percentiles. And because of this, analyzing normally distributed data is also known as parametric analysis. This contrasts with the nonparametric analysis, which deals with non-normally distributed data or data with unknown distribution patterns.

SAS's PROC UNIVARIATE procedure performs an in-depth examination of all variables by analyzing, summarizing, visualizing, and modeling the distribution of numeric variables. Outputs from the PROC UNIVARIATE procedure include (a) moments – count, mean, and standard deviation; (b) median, range, and mode; (c) custom-specific percentiles and default percentiles and quantiles; (d) extreme observations and their locations; (e) skewness and kurtosis; and (f) Q (quantile) plots and a host of other plots.

SAS's PROC UNIVARIATE also performs data distribution normality determination through four different methods: the Shapiro–Wilk test, Kolmogorov–Smirnov test, Cramer–von Mises test, and the Anderson–Darling test. The Shapiro–Wilk test is performed only when the number of observations in your dataset is fewer than 2000 (rows).

The PROC UNIVARIATE procedure offers a greater variety of descriptive statistics than the PROC MEANS procedure, which will be covered in the next section.

The basic PROC UNIVARIATE procedure is listed below along with additional examples.

```
PROC UNIVARIATE;
VAR var1 var2;
HISTOGRAM  var1 /NORMAL;
PROBPLOT var1;
QQPLOT var1 ;
RUN;
```

Example: 6.2

Data: d62and63.sas7bdat
As illustrated below, the d62and63 dataset contains vehicle speeds (13,559 rows of observation) at various times of day along State Road 40 at the location of A544 and A768.

Location	DOW	Mdate	Month	Mtime	AMspeed	Pmspeed
A768	1	1/1/17	1	5:00	42.31	37.44
A768	1	1/1/17	1	6:00	30.92	27.36
A768	1	1/1/17	1	7:00	48	42.48
A768	1	1/1/17	1	8:00	41.49	36.72
A768	1	1/1/17	1	10:00	51.26	45.36
A768	1	1/1/17	1	11:00	34.17	30.24
A768	1	1/1/17	1	12:00	47.19	41.76
A768	1	1/1/17	1	13:00	21.97	19.44
A768	1	1/1/17	1	14:00	40.68	36.23
A768	1	1/1/17	1	15:00	42.31	37.44
A768	1	1/1/17	1	16:00	35.8	31.68
A768	1	1/1/17	1	17:00	33.36	29.52
A768	1	1/1/17	1	18:00	40.68	36.00
A768	1	1/1/17	1	19:00	35.8	31.68
A768	1	1/1/17	1	20:00	39.87	35.28
A768	1	1/1/17	1	21:00	34.98	30.96
A768	1	1/1/17	1	22:00	40.68	36.00
A768	1	1/1/17	1	23:00	42.31	37.44
A768	2	1/2/17	1	0:00	39.05	34.56
A768	2	1/2/17	1	1:00	38.24	33.84

Task A: Compute descriptive statistics, including the 22nd, 57th, and 95th percentile speeds.

Task A: SAS Codes (E62a)

```
TITLE 'Descriptive Statistics only for the Speed Data';
LIBNAME mytest 'c:\sas_exercise';
PROC UNIVARIATE DATA=mytest.d62and63;
VAR AMSPEED;
output out=mytest.my_percentiles pctlpts=22 57 95 pctlpre=P;
RUN;
```

TITLE is a SAS command and declares the title is Descriptive … Data.
LIBNAME defines a SAS library named mytest representing the folder of c:\sas_exercise.
PROC UNIVARIATE DATA=mytest.d62and63 invokes the Univariate procedure for the dataset d62and63.sas7bdat in the mytest library.
VAR declares that the Univariate procedure will be run for a variable named AMSPEED.
RUN instructs SAS to execute the PROC UNIVARIATE procedure.

Task A Result (E62a):

1 ***'Descriptive Statistics only for the Speed Data'***

2 ***The UNIVARIATE Procedure***
3 ***Variable: AMspeed (AMspeed)***

4 Moments			
N 5	13559	Sum Weights 11	13559
Mean 6	38.7380972	Sum Observations 12	525249.86
Std Deviation 7	5.17081018	Variance 13	26.7372779
Skewness 8	-2.1254011	Kurtosis 14	6.94650533
Uncorrected SS 9	20709684.1	Corrected SS 15	362504.014
Coeff Variation 10	13.3481264	Std Error Mean 16	0.04440632

17 Basic Statistical Measures			
18 Location		19 Variability	
Mean 20	38.73810	Std Deviation 23	5.17081
Median 21	40.68000	Variance 24	26.73728
Mode 22	40.68000	Range 25	56.95000
		26 Interquartile Range	4.06000

27 Tests for Location: Mu0=0					
Test 28	29 Statistic		30 p Value		
31 Student's t	t 34	872.3555	Pr > |t|	<.0001	37
Sign 32	M 35	6779.5	Pr >= |M|	<.0001	38
33 Signed Rank	S 36	45965010	Pr >= |S|	<.0001	39

	Quantiles (Definition 5)	
	Quantile	Estimate
40	100% Max	61.83
41	99%	46.38
	95%	43.93
	90%	43.12
42	75% Q3	41.49
	50% Median	40.68
	25% Q1	37.43
	10%	32.54
	5%	29.29

1 – title specified.
2 – it is the default PROC UNIVARIATE procedure subtitle.
3 – variables analyzed.
4 – it refers to a set of summary parameters regarding a dataset's statistical distribution characteristics.
5 – the number of observations.
6 – the arithmetic average across all observations (N). It measures central tendency (Σ (x_i)/N).
7 – it is the standard deviation. And it characterizes dispersion. The larger the standard deviation is, the more spread out of the observations. It equals the square root of the variance ((Σ ($x_i - \bar{x})^2$/ (N – 1))$^{0.5}$).
8 – a measure of the severity of data distribution asymmetry via both the magnitude and direction. When the mean is higher than the median, it is a positive skewness. Otherwise, it is a negative skewness.
9 – it is the uncorrected sum of squares. It sums all squared observation values (Σ ($x_i)^2$).
10 – it is the coefficient of variation. It equals the ratio of the standard deviation to the mean. It is often expressed as a percentage number.
11 – it is the sum of weights used. Observations may have weights associated with them. When N = (Sum Weights), it indicates that there is no weight for any of the observations.
12 – it is the summation of all observation values (Σ (x_i)).

Quantiles (Definition 5)	
Quantile	Estimate
1%	17.09
0% Min	4.88

43

Extreme Observations			
Lowest		Highest	
Value	Obs	Value	Obs
4.88	5619	54.51	2619
4.88	4391	54.51	6336
4.88	4390	57.77	311
5.70	5620	57.77	4102
5.70	3231	61.83	503

44 45

46

Obs	P22	P57	P95
1	36.61	40.68	43.93

13 – this is the variance. It is the sum of the squared distance of each data value from the mean divided by the (number of observations –1). The formula is $\sum (x_i - \overline{x})^2/(N - 1)$.

14 – it refers to the presence and magnitude of tails associated with data distribution. A dataset with normal distribution has a kurtosis value of 0. A non-normally distributed dataset has a large positive or negative value.

15 – it is the corrected sum of squares. It equals to the sum of squared values ($\sum (x_i - \overline{x})^2$). The values are the differences of individual observations from the mean.

16 – it refers to the estimated standard deviation of the sample means. And it is computed as the sample standard deviation divided by the square root of the sample size.

17 – it refers to several parameters computed by the procedure.

18 – it refers to several location parameters such as the mean, median, and mode as listed.

19 – it refers to several dispersion parameters.

20 – it refers to the arithmetic average across all observations ($\sum (x_i)/N$). It measures central tendency.

21 – it refers to the middle value among all ordered observations. It characterizes the central tendency.

22 – it refers to the most occurred observations among all observations. The mode value has the highest probability to occur.

23 – it is the standard deviation. It equals to the square root of the variance $(\sum (x_i - \overline{x})^2/(N - 1))^{0.5}$. It characterizes dispersion. The larger the standard deviation is, the more spread out of the observations.

24 – it is the sum of the squared distance of each data value from the mean divided by the (number of observations –1). The formula is $(\sum (x_i - \overline{x})^2)/(N - 1)$.

25 – it refers to the difference between the largest and smallest observations among all observations. It is a parameter characterizing dispersion measurement.

26 – it measures the difference between the upper and lower quartiles.

27 – it is the statistical testing on whether the mean or mode equals to 0 (the location of mean or mode). SAS performs three different location tests. Student's t-test is appropriate to data which has a normal distribution (mean testing). It computes the value of t which is listed under the title of Statistics and the p value based on the calculated t value. The Sign test and the Signed Rank test are nonparametric tests for non-normally distributed data (median, the 50th percentile testing). The Sign test calculated a value named M, where p value can be inferred from. The Signed Rank test calculates a value named S, where p value can be inferred from. For the data shown, all the p values are smaller than 0.0001. Consequently, all three null hypotheses are rejected regardless of data distribution types.

28 – the types of statistical tests performed.

29 – the exact statistical parameters computed.

30 – it is the p value derived. Accepting or rejecting the null hypothesis relies on the p value.

31, 32, 33, 34, 35, 36, 37, 38, and 39 – see 27 for answers.

40 – 100% of all observations are equal to or less than 61.83 (maximum value).

41 – 99th percentile meaning 99% of the observations are equal to or less than 46.38.

42 – 75th percentile or the 3rd quantile meaning 75% of the observations are equal to or less than 41.49.

43 – 0% of all values are less than 4.88 (minimum value).

44 – the lowest 5 and highest 5 observations. 4.88 is the lowest value, and it is from row number (Obs) 5619.

45 – 61.83 is the highest value, and it is from row number (Obs) 503.

46 – the 22nd percentile value is 36.61.

Task B: Create a Histogram

Task B: SAS Codes (E62b)

```
TITLE 'Histogram Only
for the Speed Data';
LIBNAME mytest 'c:\
sas_exercise';
PROC UNIVARIATE
DATA=mytest.d62and63
NOPRINT;
HISTOGRAM;
VAR PMSPEED;
RUN;
```

TITLE declares that "Histogram Only for the Speed Data" is the title for all outputs.

LIBNAME defines a SAS library named mytest representing the folder of c:\sas_exercise.

PROC UNIVARIATE DATA=mytest.d62and63 NOPRINT invokes the Univariate statistical procedure for the dataset d62and63.sas7bdat in the library of mytest. The NOPRINT option instructs SAS not to print any analysis result.

HISTOGRAM is a SAS command requesting SAS to generate a histogram.

VAR declares that the variable PMSPEED is to be analyzed by the Univariate procedure.

RUN instructs SAS to execute the PROC UNIVARIATE procedure.

Task B Result (E62b):

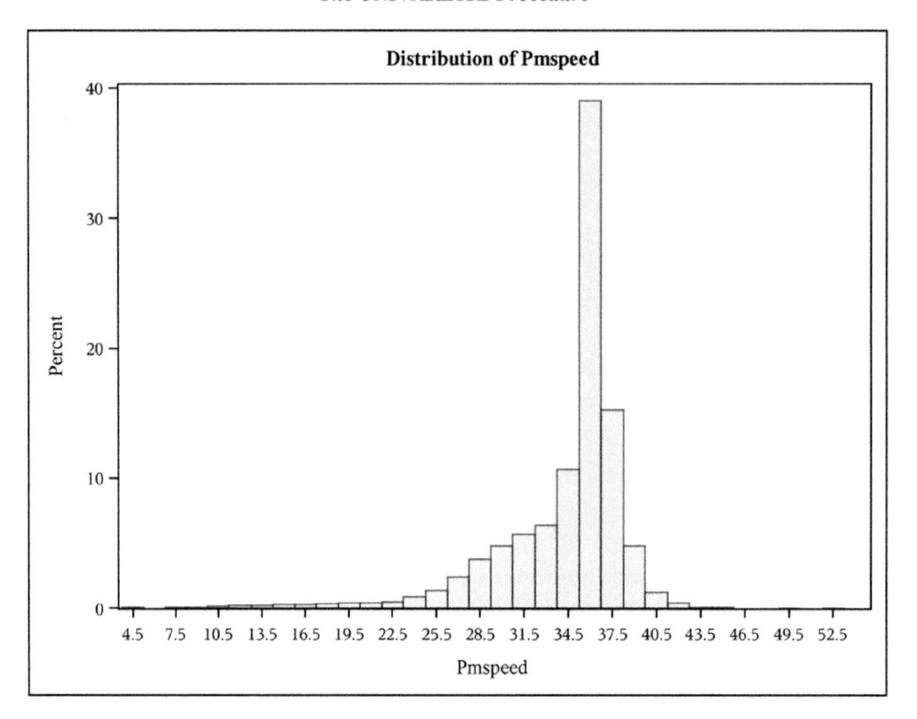

The diagram is a histogram for the PMspeed data plot. The horizontal axis is the PMspeed grouped in 3 mph increments with their middle points plotted. The Y-axis is the percentage of observations each speed block represents.

Clearly, the distribution shows the data is left-skewed and does not resemble a normal distribution pattern.

Task C: Normality Testing

Task C: SAS Codes (E62c)

```
TITLE 'Speed Data Distribution (Normality) Test';
LIBNAME mytest 'c:\sas_exercise';
PROC UNIVARIATE DATA=mytest.d62and63 NORMAL PLOT;
QQPLOT;
VAR AMspeed;

RUN;
```

TITLE declares that Speed Data Distribution (Normality) Test is the title for all outputs.
LIBNAME defines a library named mytest representing the folder of c:\sas_exercise.
PROC UNIVARIATE DATA=mytest.d62and63 NORMAL PLOT invokes the Univariate statistical procedure for d62and63.sas7bdat data in the mytest library.
NORMAL PLOT is an option. The keyword NORMAL asks SAS to perform a data distribution normality test with four methods deployed by SAS. The PLOT option instructs SAS to plot a probability (frequency) chart for variables specified. Also, the PLOT option instructs SAS to perform a quantile plot for variables specified.
VAR specifies that the variable to be analyzed with the Univariate procedure is AMspeed.
RUN instructs SAS to execute the PROC UNIVARIATE procedure.

Task C Result (E62c):

1 ***'Speed Data Distribution (Normality) Test'***

2 ***The UNIVARIATE Procedure***
3 ***Variable: AMspeed (AMspeed)***

4 Moments			
N 5	13559	Sum Weights 11	13559
Mean 6	38.7380972	Sum Observations 12	525249.86
Std Deviation 7	5.17081018	Variance 13	26.7372779
Skewness 8	-2.1254011	Kurtosis 14	6.94650533
Uncorrected SS 9	20709684.1	Corrected SS 15	362504.014
Coeff Variation 10	13.3481264	Std Error Mean 16	0.04440632

17 Basic Statistical Measures			
18 Location		19 Variability	
Mean 20	38.73810	Std Deviation 23	5.17081
Median 21	40.68000	Variance 24	26.73728
Mode 22	40.68000	Range 25	56.95000
	26	Interquartile Range	4.06000

27 Tests for Location: Mu0=0					
Test 28	29 Statistic		30 p Value		
31 Student's t	t 34	872.3555	Pr > |t|	<.0001	37
Sign 32	M 35	6779.5	Pr >= |M|	<.0001	38
33 Signed Rank	S 36	45965010	Pr >= |S|	<.0001	39

40 Tests for Normality				
Test 41	42 Statistic		43 p Value	
Kolmogorov-Smirnov 44	D	0.196195	Pr > D 47	<0.0100
Cramer-von Mises 45	W-Sq	131.692	Pr > W-Sq 48	<0.0050
Anderson-Darling 46	A-Sq	699.5006	Pr > A-Sq 49	<0.0050

The interpretation of all the data items (1–39) follows the same definitions used in Task A Result (E62a) except the Test for Normality (40–49) information as explained below.
40 – statistical analysis in determining whether the data has a normal distribution.
41 – under this title, three different tests are carried out.
42 – type of statistical values computed for various analysis methods.
43 – p values derived based on the corresponding statistical value derived in 42
44 – Kolmogorov–Smirnov method.
45 – Cramer–von Mises method.
46 – Anderson–Darling method.
These methods generate the statistical values of D, W-Sq, and A-Sq, respectively. Based on the D, W-Sq, and A-Sq values computed, their corresponding p values (47, 49, and 49) are determined.
For this example, the Shapiro–Wilk test is not used because the data has more than 2000 observations (rows). Based on the p values (all are smaller than 0.5%) of all the three tests, the hypothesis for the "Tests for Normality" is rejected. The data does not have a normal distribution.

The UNIVARIATE Procedure

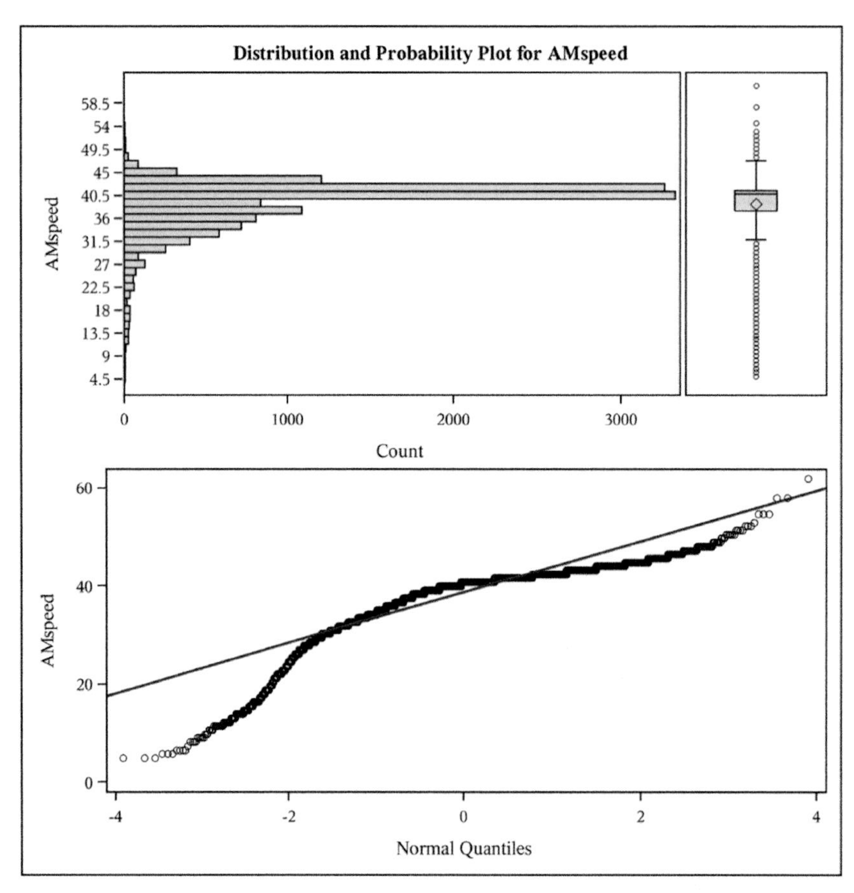

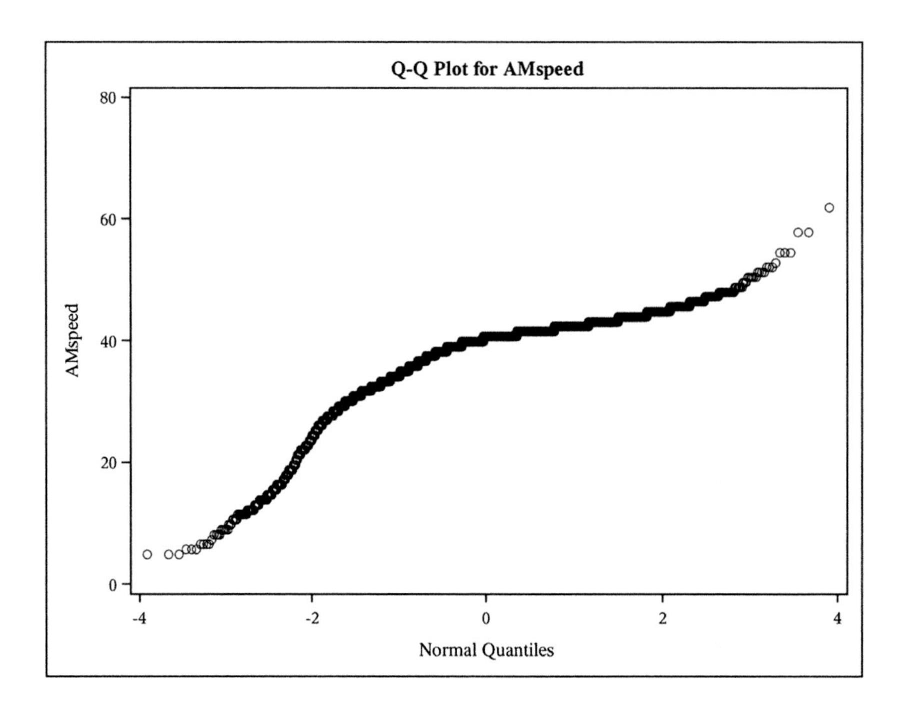

The top part of the "Distribution and Probability Plot for AMspeed" is the AMspeed distribution plot. The horizontal axis has the number of observations of the speed in each of the speed bins identified in the Y-axis. Obviously, the AMspeed data is highly skewed towards the low-speed tail.

The bottom chart is the probability plot with an assumed normal distribution pattern characterized by the straight solid line. The horizontal axis with the subtitle of Normal Quantiles follows the normal distribution layout for ordered observations (…, $\mu - 3s$, $\mu - 2s$, $\mu - s$, $\mu = 0$, $\mu + s$, $\mu + 2s$, $\mu + 3s$, …). The thick dotted line represents the actual AMspeed data probability for the speed identified on the Y-axis. The thin straight solid upward line represents the speed data probability if the data is normally distributed. Clearly, the thick line does not follow the straight line, which is a clear indication that the data is not normally distributed.

The Q–Q plot is AMspeed, which is the same as the probability plot except no normal plot drawing (the straight line) is provided.

6.3 MEAN AND DESCRIPTIVE STATISTICS – PROC MEANS

The PROC MEANS procedure produces descriptive statistics for variables across all observations and within groups of observations. When you don't need the more extensive descriptive statistics generated by PROC UNIVARIATE, you can use PROC MEANS to achieve your goal.

The general setup of the PROC MEANS procedure is listed below:

```
LIBNAME mydata 'c:\sas_exercise';
PROC MEANS DATA=mydata.d62and63;
VAR val var1;
RUN;
```

As with most of the PROC procedures, options are available for the PROC MEANS procedure as listed below:

PROC Means Options:

- CSS – corrected sum of squares
- MAX – maximum (largest)
- MEAN – arithmetic average)
- MIN – minimum (smallest)
- N – number of observations
- NMISS – number of missing observations
- PRT – p value associated with t-test above
- RANGE – range
- STD – standard deviation
- STDERR – standard error
- SUM – sum of observations

- SUMWGT – sum of the WEIGHT variable values
- T – Student's t value for testing Ho: md = 0
- USS – uncorrected sum of squares
- VAR – variance
- MEDIAN – 50th percentile
- P1 – 1st percentile
- P5 – 5th percentile
- P10 – 10th percentile
- P90 – 90th percentile
- P95 – 95th percentile
- P99 – 99th percentile
- Q1 – 1st quartile
- Q3 – 3rd quartile
- QRANGE – quartile range.

Example: 6.3

Task A: Use the PROC MEANS without Specifying Options

Task A: SAS Codes (E63a)

```
TITLE " PROC Means - Speed
Data";
LIBNAME mym 'c:\sas_exercise';
PROC MEANS DATA=mym.d62and63 ;
VAR AMspeed PMspeed;
RUN;
```

TITLE declares that "PROC Means - Speed Data" is the title for all outputs.

LIBNAME defines a library named mym representing the folder of c:\sas_exercise.

PROC MEANS DATA invokes the means statistical procedure to analyze the data named d62and63.sas7bdat in the library of mym.

VAR tells SAS that variables AMspeed and PMspeed are to be analyzed with the Means procedure.

RUN instructs SAS to execute the PROC MEANS procedure.

Task A Result (E63a):

1 ***PROC Means - Speed8a Data***

2 ***The MEANS Procedure***

3	4	5	6	7	8	9
Variable	**Label**	**N**	**Mean**	**Std Dev**	**Minimum**	**Maximum**
Amspeed	Amspeed	5707	55.1869634	6.0562776	5.0000000	62.0000000
Pmspeed	Pmspeed	5707	49.9123883	5.4812588	4.0000000	56.0000000

1 – title specified.
2 – it is the default PROC MEANS subtitle.
3 – Amspeed and Pmspeed variables.
4 – the variable name is the same as the label meaning no label command is used.
5 – the number of observations for each variable.
6 – the arithmetic means of all observations ($\sum (x_i)$)/N.
7 – it is the standard deviation. It equals to the square root of the variance.
8 – the smallest observed value.
9 – the largest observed value.

Task B: Additional Options Specified

Task B: SAS Codes (E63b)

```
TITLE " PROC Means - Speed Data";
LIBNAME mym 'c:\sas_exercise';
PROC MEANS DATA=mym.d62and63  N NMISS MEAN Range P5 P95;
VAR AMspeed PMspeed;
RUN;
```

TITLE declares that "PROC Means -Speed Data" is the title to be used with all outputs.
LIBNAME defines a library named mym representing the folder of c:\sas_exercise.
PROC MEANS DATA=mym.d62and63 invokes the means statistical procedure to analyze the data named d62and63.sas7bdat in the library of mym. The N (the number of valid observations), NMISS (the number of observations has missing values), MEAN, Range, P5 (5th percentile value), and P95 (95th percentile value) are options specified with the Means procedure.
VAR tells SAS that the variables AMspeed and PMspeed are the ones from the d62and63 data that need to be analyzed.
RUN instructs SAS to execute the PROC MEANS procedure.

Task B Result (E63b):

1 ***PROC Means - Speed8a Data***

2 ***The MEANS Procedure***

3 Variable	4 Label	5 N	6 N Miss	7 Mean	8 Range	9 5th Pctl	10 95th Pctl
Amspeed	Amspeed	5707	3	55.1869634	57.0000000	43.0000000	59.0000000
Pmspeed	Pmspeed	5707	3	49.9123883	52.0000000	39.0000000	53.0000000

1 – it is the title specified.
2 – it is the default PROC MEANS subtitle.
3 – Amspeed and Pmspeed are the variables analyzed.
4 – it is a label. You can use the Label command to display a variable with a different name. Here no label commands are used. Variables maintain their original names.
5 – the number of observations for each variable.
6 – the number of observations with no entry.
7 – the arithmetic means of all observations (Σ (x_i)/N).
8 – the difference between the largest and the smallest observations.
9 – the 5th percentile value.
10 – the 95th percentile value.

Task C: Using the BY and CLASS Statement with PROC Means
When you use the "BY var" statement with the PROC MEANS, each "BY group" will be analyzed separately by the PROC MEANS procedure, and a different output data table is generated for each of the "BY group."

SAS also has the "CLASS var" statement in dealing with categorical variables (discrete numerical variable, groups of continuous variables). If you use the "CLASS var" instead of the "By var" in the PROC MEANS procedure, one table containing all the classification variables is generated.

For example, the variable "day of week" (DOW) has a value from 1 to 7. When you use the "BY DOW" statement, speeds for each day of the week are analyzed separately, and seven different tables are generated. When you use the "CLASS DOW" statement, speeds for DOW 1 to 7 are still analyzed separately, but the result will be presented in a single table.

Example: 6.3c

Case 1: Use the BY Statement (BY DOW which is Day 1 to Day 7)

Case 1: SAS Codes (E63cl)

```
TITLE " PROC Means BY DOW - Speed Data";
LIBNAME mym 'c:\sas_exercise';
PROC SORT data=mym.d62and63;
BY DOW;
RUN;
PROC MEANS DATA=mym.d62and63 ;
VAR AMspeed;
BY DOW;
RUN;
```

TITLE declares that "PROC Means by DOW - Speed Data" is the title to be used with all outputs.
LIBNAME defines a library named mym representing the folder of c:\sas_exercise.
PROC SORT data=mym.d62and63; invokes the sort command to sort a dataset named d62and63.ass7bdat in the mym library.

BY DOW tells SAS that sorting is based on the variable of DOW.
RUN; kicks off the PROC SORT procedure.
PROC MEANS DATA=mym.d62and63 invokes the means procedure for data called d62and63.sas7bdat in the mym library.
VAR tells SAS to carry out the means procedure for the variable of AMspeed.
BY DOW; tells SAS that the means procedure for the variable of AMspeed is by DOW and should be run per DOW.
RUN instructs SAS to execute the PROC MEANS procedure.

Case 1 Result (E63c1):

1 ***PROC Means BY DOW - Speed Data***

2 ***The MEANS Procedure***

3 **DOW=1**

4 Analysis Variable : AMspeed AMspeed				
5 N	6 Mean	7 Std Dev	8 Minimum	9 Maximum
1888	40.0360434	4.6502896	11.3900000	61.8300000

DOW=2

Analysis Variable : AMspeed AMspeed				
N	Mean	Std Dev	Minimum	Maximum
1954	38.4859161	5.1512025	5.7000000	54.5100000

DOW=3

Analysis Variable : AMspeed AMspeed				
N	Mean	Std Dev	Minimum	Maximum
1953	38.3226421	5.2951887	4.8800000	57.7700000

DOW=4

Analysis Variable : AMspeed AMspeed				
N	Mean	Std Dev	Minimum	Maximum
1956	38.2252761	5.4991657	8.1400000	54.5100000

DOW=5

Analysis Variable : AMspeed AMspeed				
N	Mean	Std Dev	Minimum	Maximum
1959	38.2295048	5.4779673	6.5100000	48.8200000

DOW=6

Analysis Variable : AMspeed AMspeed				
N	Mean	Std Dev	Minimum	Maximum
1941	38.3302524	5.1414513	6.5100000	52.8800000

1 – it is the title specified.
2 – default PROC MEANS subtitle.
3 – analysis for the DOW=1 period resulted from the BY DOW statement.
4 – AMspeed is the variable to be analyzed.
5 – the number of observations for each variable.
6 – the arithmetic means of all observations.
7 – it is the standard deviation. It equals to the square root of the variance.
8 – the smallest observed value.
9 – the largest observed value.
Results from all other DOW periods follow the same pattern.

Case 2: Use the Class Statement (Class DOW which is Day 1 to Day 6)

Case 2: SAS Codes (E63c2)

```
TITLE " PROC Means BY DOW - Speed Data";
LIBNAME mym 'c:\sas_exercise';
PROC MEANS DATA=mym.d62and63 ;
VAR AMspeed;
CLASS DOW;
RUN;
```

TITLE declares that "PROC Means BY DOW - Speed Data" is the title to be used with all outputs.
LIBNAME defines a library named mym representing the folder of c:\sas_exercise.
PROC SORT data=mym.d62and63; invokes the sort command to sort a dataset named d62and63.ass7bdat in the mym library.
BY DOW tells SAS that sorting is based on the variable of DOW.
RUN kick-starts the PROC SORT procedure.
PROC MEANS DATA=mym.d62and63 invokes the means procedure for data called d62and63.sas7bdat in the mym library.
VAR instructs SAS to carry out the means procedure for the variable AMspeed.
CLASS declares that the variable DOW is to be treated as a categorical variable.
RUN instructs SAS to execute the PROC MEANS procedure.

Case 2 Result (E63c2):

1 ***PROC Means BY DOW - Speed8a Data***

2 ***The MEANS Procedure***

3 Analysis Variable : Amspeed Amspeed						
4 DOW	5 N Obs	6 N	7 Mean	8 Std Dev	9 Minimum	10 Maximum
1	815	815	56.4024540	5.4020193	10.0000000	62.0000000
2	815	815	55.4110429	5.6578641	8.0000000	61.0000000
3	813	813	55.5338253	5.6788346	5.0000000	61.0000000
4	816	816	55.3799020	5.6765831	7.0000000	61.0000000
5	816	816	54.3860294	6.3144008	8.0000000	61.0000000
6	816	816	53.3713235	7.4024112	7.0000000	60.0000000
7	816	816	55.8272059	5.5332388	18.0000000	62.0000000

1 – it is the title specified.
2 – it is the default PROC MEANS subtitle.
3 – Amspeed is the variable to be analyzed.
4 – all periods from 1 to 7 resulting from the Class DOW statement.
5 – the number of observations with no missing values for each variable.
6 – the number of observations, including missing values for each variable.
7 – the arithmetic means of all observations $(\sum (x_i))/N$.
8 – it is the standard deviation. It equals to the square root of the variance.
9 – the smallest observed value.
10 – the largest observed value.

6.4 DETERMINING THE DIFFERENCE BETWEEN TWO MEANS: PROC TTEST

Student's t-test compares two sets of sample data to decipher whether there is any statistically significant difference between them in regard to the population means they represent.

There are three types of Student's t-test and one special case. The first type of Student's t-test is the one-sample t-test, where the mean of the sample data is compared with a known value. The second type of Student's t-test is the paired t-test, where observations from paired samples are compared. If possible, paired sampling (during experimental design and data collection) should be considered first, as it offers better control in quantifying differences between the two sample groups.

The third type of Student's t-test is the independent sample t-test, where means from two independently collected samples are compared. You will likely encounter this third type of t-test in most situations.

The paired and the independent t-test methods assume that the variances of the two sample groups you are testing are equal in the population they come from.

In the event that the equal variance assumption cannot be ascertained, a modified method known as Welch's t-test should be used. In SAS, Welch's t-test is an option under the name "Satterthwaite."

Example 6.4.1: Paired dataset

Data: tspeed.xlsx
The tspeed data has three variables: location, loop speed, and Bluetooth speed. At a given location, truck speed is measured by both the newly installed Bluetooth sensors and the existing loop sensors. The data layout is illustrated below.

location	loop (mph)	bluetooth (mph)
pg12b	45.4	44.8
pg12b	36.2	34.5
pg12b	34.1	34.2
pg12b	48.1	47.9
pg12b	54.2	55.1
pg12b	51.6	50.7
pg12b	17.5	18.3
pg12b	49.4	48.1
pg202	56.8	60.1
pg202	65.9	66.7
pg202	62.3	63.1
pg202	63.6	63.5
pg202	45.4	46.6
pg202	56.7	56.9
pg202	67.6	68.9
ar666	12.3	11.6
ar666	6.9	7.2
ar666	11.2	12.5

Task: Determine whether the two measuring methods (loop and Bluetooth) have any difference with regard to speed measurement.

Analysis: The paired Student's t-test is the right tool for this analysis. It is a two-mean comparison based on samples from paired measurements. We can analyze all observations across all locations, as illustrated in Case A below. We can also analyze the paired data by location with a "BY location;" statement, as illustrated in Case B.

Case A: SAS Codes to Analyze all Observations Regardless of Locations

Case A: SAS Codes (E641a)

```
TITLE "PROC TTtest - Paired Method for Speed measurement";
LIBNAME myp 'c:\sas_exercise';

PROC IMPORT OUT= WORK.tpair
            DATAFILE= "c:\sas_
exercise\tpair.xlsx"
            DBMS=EXCEL REPLACE;
     RANGE="Sheet1$";
     GETNAMES=YES;
     MIXED=YES;
     SCANTEXT=YES;
     USEDATE=YES;
     SCANTIME=YES;
RUN;

PROC TTEST DATA=tpair ;
PAIRED LOOP * Bluetooth;
RUN;
```

Explanations for codes from the TITLE statement until the PROC TTEST statement can be obtained from Section 3.5.1 Importing an Excel File.

PROC TTEST DATA=tpair; invokes SAS's t-test procedure for a dataset named tpair.sas7bdat in the WORK library.

*PAIRED LOOP * Bluetooth specifies paired observations which are represented by the variables loop and Bluetooth. Keep in mind that each row is a pair of observations for the same event with the two methods. The * symbol between loop and Bluetooth is used to indicate the data are in pairs.*

RUN instructs SAS to execute the PROC TTEST procedure.

Case A Result (E641a):

1 PROC TTtest - Paired Method for Speed measurement

2 The TTEST Procedure

3 Difference: loop - bluetooth

4 N	5 Mean	6 Std Dev	7 Std Err	8 Minimum	9 Maximum
20	-0.3550	1.1381	0.2545	-3.3000	1.7000

10 Mean	11 95% CL Mean		12 Std Dev	13 95% CL Std Dev	
-0.3550	-0.8876	0.1776	1.1381	0.8655	1.6623

14 DF	15 t Value	16 Pr > |t|
19	-1.39	0.1791

1 – it is the title specified.

2 – it is the default subtitle of the PROC TTEST.

3 – the parameter to be tested with a paired t-test is the difference between loop and Bluetooth.

4 – 20 is the number of paired observations (20 rows).

5 – it is the arithmetic means of the differences between the paired loop and Bluetooth values $(\sum (x_i - z_i))/N$.

6 – it is the standard deviation. It is the square root of the variance. It measures central tendency.

7 – the standard error measures how far a sample mean is from the population mean. It is obtained by dividing the Std Dev with the square root of the observation number (N). In this example, it is $0.2545 = 1.1381/\sqrt{20}$.

8 – the smallest observed value.

9 – the largest observed value.

10 – the arithmetic means of the differences between loop and Bluetooth paired values. In this case, it is –0.3550.

11 – it is the 95% confidence bounds for the mean. It has two values representing the upper and lower 95% confidence bounds for the mean. The –0.8876 is the upper bound, and 0.1381 is the lower bound.

12 – it is the standard deviation, and it equals the square root of the variance. It measures central tendency.

13 – it is the 95% confidence bounds for the standard deviation. It has two values representing the standard deviations for the upper and lower confidence bound means. The 0.8655 is the upper bound mean standard deviation, and 1.6623 is the lower bound mean standard deviation.

14 – it is the degree of freedom.

15 – it is the t value calculated and equals to the Mean divided by the Std Err. Here, it is −0.3550/0.2545 = −1.3948.

16 – it is the two-tailed probability computed using the t-distribution based on the DF and t value. In this example, the p value is 0.1791. The hypothesis that there is no difference between the Bluetooth and loop paired values is failed to be rejected. It is concluded that there is no statistically significant difference between the two measurement methods with regard to speed determination.

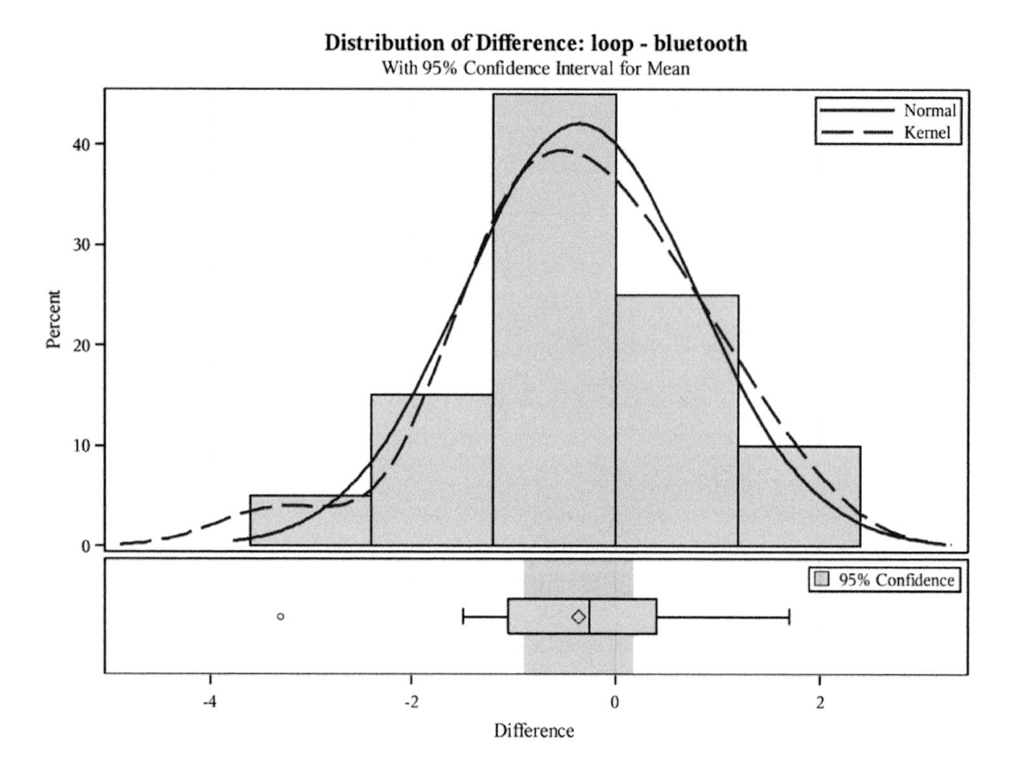

The Distribution of Difference: loop - bluetooth – the top part overlays three sets of information. The bar chart is a frequency histogram for the mean differences. The solid line represents a normally distributed data distribution pattern. The dashed line is the so-called kernel estimates of the percentiles of the hypothesized probability density. The kernel used is the Gaussian density with the assumed variance based on the observed data. The kernel density determines the smoothness of the resulting estimate. The bottom part of the chart is the box plot of the data and the 95% confidence intervals. In this case, the kernel density curve (dashed line) approximates the bar chart, which has a great similarity to the normal density curve (solid line).

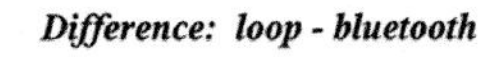

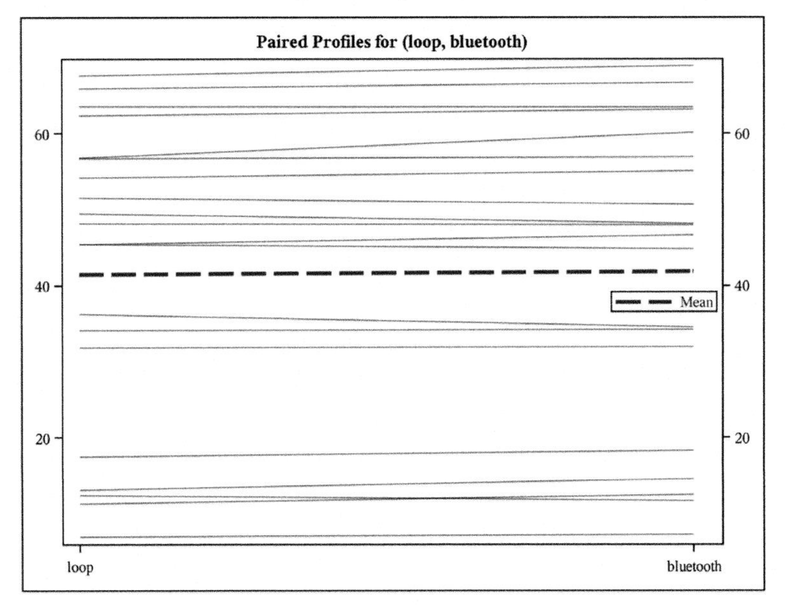

The Paired Profiles for (loop, bluetooth) chart is a simple line chart connecting paired measurements with loop values on the left vertical axis and Bluetooth values on the right vertical axis. All the horizontal lines have different slopes. Solid lines represent the actual observation pairs. The dashed line represents the means. This diagram provides a visual aid in deciphering the difference (steepness of the slops) between the loop and Bluetooth values.

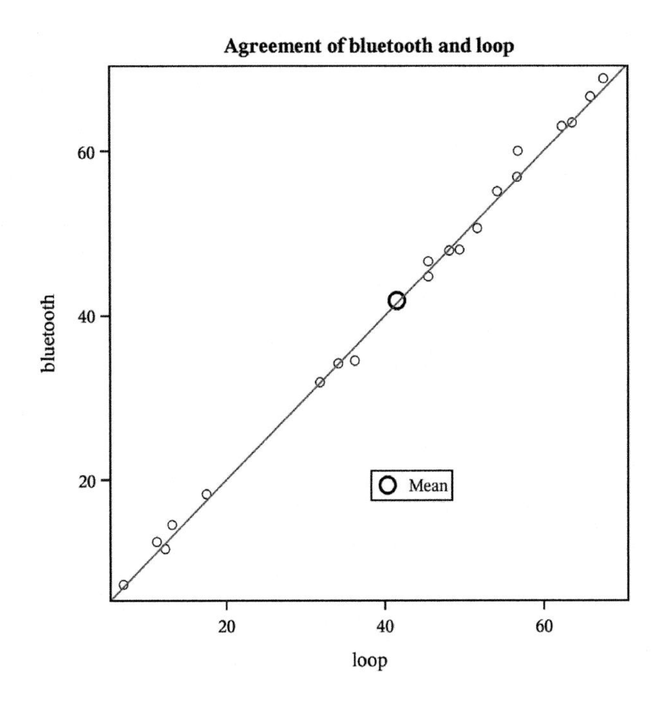

The Agreement of Bluetooth and loop chart offers another perspective on the relationships between the loop and Bluetooth values. Virtually all data points are centered in pairs along the diagonal equal line. In this case, it is a sign indicating the closeness of the paired data.

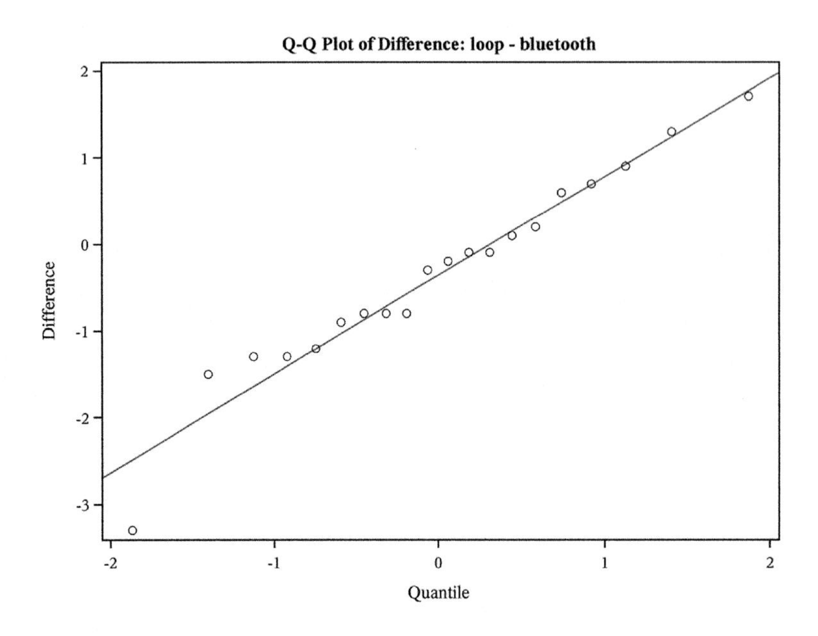

The Q–Q Plot of Difference: loop - Bluetooth chart shows the data distribution pattern. The horizontal axis is the quantile value, and the vertical axis represents the difference between a pair of observations. The solid diagonal line represents data distribution following a perfect normal Gaussian pattern. The empty circle represents the actual observed data value differences which closely follow the solid line, indicating its normal distribution characteristics.

Case B: SAS Codes to Analyze Observations by Locations

Case B: SAS Codes (E641b)

```
TITLE "PROC TTest-Paired
Method by Location";
LIBNAME myp 'c:\sas_exercise';

PROC IMPORT OUT= WORK.tpair
            DATAFILE= "c:\
sas_exercise\tpair.xlsx"
            DBMS=EXCEL REPLACE;
    RANGE="Sheet1$";
    GETNAMES=YES;
    MIXED=NO;
```

Explanations for codes from the TITLE statement until the PROC SORT statement can be obtained from Section 3.5.1 Importing an Excel File.

PROC SORT DATA=tpair BY location invokes the sort command to sort data tpair.sas7bdat data in the WORK library by location.

PROC TTEST DATA=tpair invokes SAS's t-test procedure for data named tpair in the WORK library.

```
        SCANTEXT=YES;
        USEDATE=YES;
        SCANTIME=YES;
RUN;

PROC SORT DATA=tpair;
BY location;
RUN;

PROC TTEST DATA=tpair ;
PAIRED LOOP * Bluetooth;
BY Location;
RUN;
```

*PAIRED LOOP * Bluetooth; specifies that the paired observations are represented by variables of loop and Bluetooth. Keep in mind that each row is a pair of observations. The * symbol between loop and Bluetooth is used to indicate the data are in pairs.*

BY Location directs SAS to carry out the t-test for each location independently.

RUN instructs SAS to execute the PROC TTEST procedure.

Case B Result (E641b):
The output from E641b follows the same format of E641a for each of its locations and is omitted here.

Example 6.4.2: Independent Two Samples

Data: tguard.xlsx
The tguard data has two variables: device and performance. A total of 61 measurements ((29 for BCC26 and 32 for KSS166)) were carried out on two different guardrails (BCC26 and KSS166)after they were crashed by vehicles. Each crashed device was given a rating of 1–10 on its performance. The data layout is illustrated below.

Device	Performance
BCC26	8
BCC26	8
BCC26	6
BCC26	9
KSS166	6
KSS166	5
KSS166	7
KSS166	8

Task: Assess whether there is any difference between the performances of the two different guardrail devices.

Analysis: As described, the data are two independent samples. The standard t-test applies to determine the mean difference.

SAS Codes (E642)

```
TITLE "PROC TTtest-Independent Samples";
LIBNAME myp 'c:\sas_exercise';
```

```
PROC IMPORT OUT= WORK.tguard
            DATAFILE= "c:\sas_
exercise\tguard.xlsx"
            DBMS=EXCEL REPLACE;
     RANGE="Sheet1$";
     GETNAMES=YES;
     MIXED=NO;
     SCANTEXT=YES;
     USEDATE=YES;
     SCANTIME=YES;
RUN;

PROC TTEST DATA=tguard ;
CLASS device;
VAR performance;
RUN;
```

Explanations for codes from the TITLE statement until the PROC TTEST statement can be obtained from Section 3.5.1 Importing an Excel File.

PROC TTEST DATA=tguard invokes SAS's t-test procedure for a dataset named tguard in the WORK library.

CLASS specifies that the device (Bluetooth and Loop) is a categorical variable.

VAR performance tells SAS that the variable to be analyzed is performance.

RUN instructs SAS to execute the PROC TTEST procedure.

Result (E642):

1 ***PROC TTest-Independent Samples***

2 ***The TTEST Procedure***

3 ***Variable: Performance (Performance)***

	Device 4	N 5	Mean 6	Std Dev 6	Std Err 7	Minimum 8	Maximum 9
10	BCC26	29	5.7931	2.3812	0.4422	1.0000	9.0000
11	KSS166	32	5.6875	2.7408	0.4845	1.0000	9.0000
12	Diff (1-2)		0.1056	2.5764	0.6605		

13 Device	14 Method	15 Mean	16 95% CL Mean		17 Std Dev	18 95% CL Std Dev	
BCC26		5.7931	4.8874	6.6989	2.3812	1.8896	3.2204
KSS166		5.6875	4.6993	6.6757	2.7408	2.1973	3.6439
Diff (1-2)	Pooled	0.1056	-1.2162	1.4274	2.5764	2.1838	3.1423
Diff (1-2)	Satterthwaite	0.1056	-1.2070	1.4182			

	Method	Variances 19	DF 20	t Value 21	Pr > |t| 22
23					
24	Pooled	Equal	59	0.16	0.8735
25	Satterthwaite	Unequal	58.904	0.16	0.8726

26 Equality of Variances

27 Method	Num DF 28	Den DF 29	F Value 30	Pr > F 31
Folded F	31	28	1.32	0.4549

1 – it is the title specified.

2 – it is the default subtitle of the PROC TTEST.

3 – it shows that the test is for the variable Performance.

4 – variable name is Device. It has two different values in this case.

N – the number of observations (rows): 29 for BCC26 and 32 for KSS166.

5 – it is the arithmetic means for each of the device values $(\sum (x_i))/N$*.*

6 – it is the standard deviation. It is the square root of the variance. It measures central tendency.

7 – it is the standard error, and it measures how far a sample mean is from the population mean. It is obtained by dividing the Std Dev with the square root of the observation numbers (N).

8 – the smallest observed value.

9 – the largest observed value.

10 – the name of one observation.

11 – the name of the other observation.

12 – it refers to the simple difference between the two observations for the Mean, Std Dev, and Std Err between BCC26 and KSS166.

13 – variable name. It has two different values in this case.

14 – it refers to the relevant computation based on the variance assumption (equal or not equal) used.

15 – it is the arithmetic means for the appropriate observations.

16 – it refers to the upper and lower bounds of the 95% mean confidence interval.

17 – it is the standard deviation and equals to the square root of the variance.

18 – it refers to the 95% confidence levels (upper and lower bounds) for the standard deviation.

Diff (1–2) Pooled Method – if the two populations represented by BCC26 and KSS166 have the same variance, the Pooled variance method is the one to be relied on. Otherwise, the Satterthwaite method is more appropriate.

19 – it is the variance computed with the Pooled or Satterthwaite method.

20 – degree of freedom.

21 – it is the t value and equals to the Mean divided by the Std Err.

22 – it is the two-tailed probability computed using the t-distribution based on the DF and t value.

24 – the p value is 0.8735 under the Pooled method

25 – the p value is 0.8726 under the Satterthwaite method.

The hypothesis that there is no difference between the performance of the BCC26 and KSS166 is failed to be rejected based on 24 and 25. It is concluded that there is no statistically significant difference between the devices in terms of their performance regardless of the data variance equality assumption issue.

26 – this table addresses the issue of which t-test (Pooled variance or Satterthwaite variance) is more appropriate.

27 – Folded F is the method used. It refers to the F test (variance) by always using the larger variance as the numerator and the smaller variance as the denominator between the two tested samples. In this example, the two samples under the variable Device are the BCC26 and KSS166. BCC26 has an Std Dev of 2.3812, and KSS166 has an Std Dev of 2.7408. The $F = 2.7408^2 / 2.3812^2 = 1.32$*.*

28 – the degree of freedom for the numerator. Here, the numerator is the KSS166. KSS166 variance is larger than the BCC26 variance.

29 – the degree of freedom for the denominator. Here, the denominator is the BCC26 because its variance is smaller than the KSS166 sample variance.

30 – this is the F statistics and is calculated by dividing the larger variance by the smaller variance [max(var1 and var2)/min (var2 and var2)].

31 – it refers to the F test. In this example, the p value is greater than 0.4549, which indicates that the hypothesis (equal variance) is failed to be rejected. There is no statistically significant difference between the two variances.

Based on this conclusion, the final conclusion on the t-test should be based on the Pooled variance method. And the conclusion is that there is no statistically significant difference between the two devices in terms of their performances (p value = 0.8735).

PROC TTtest-Independent Samples

The TTEST Procedure

Variable: Performance (Performance)

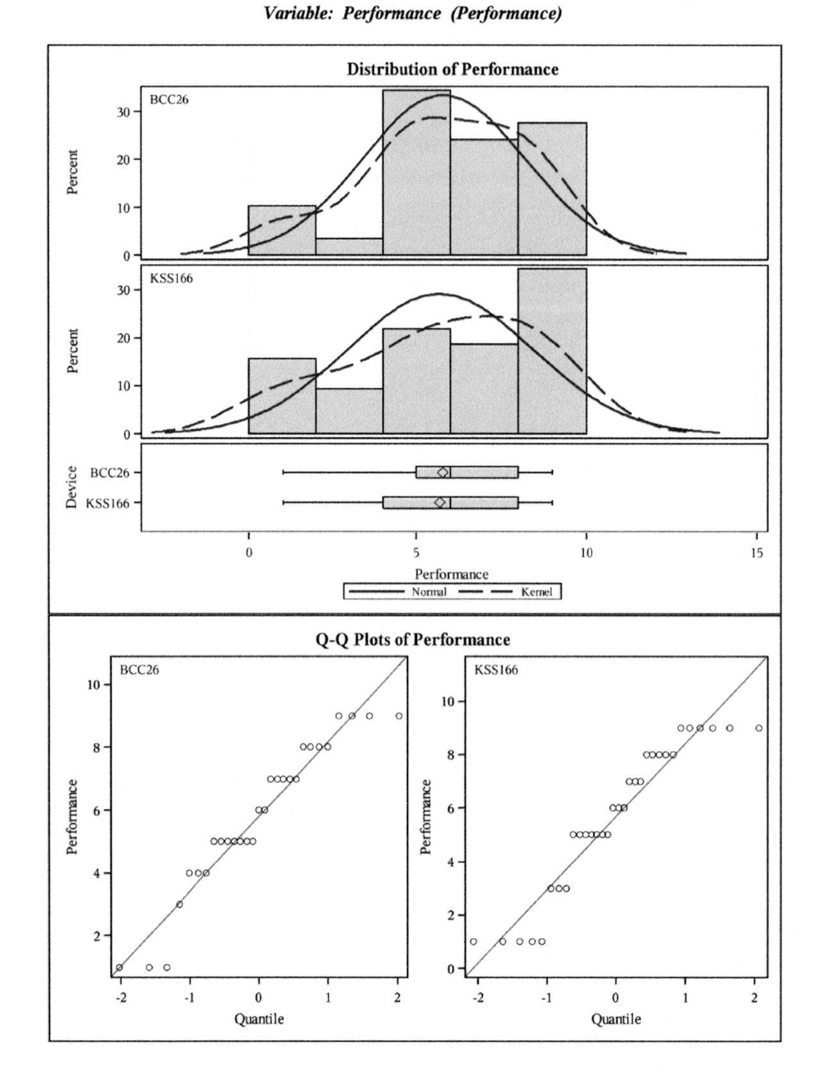

The Distribution of Performance – the top part overlays three sets of information for BCC26 and KSS166, respectively. The bar chart shows the frequency distribution for all the observations. The solid line represents a normally distributed data distribution. The dashed line is the so-called kernel estimates of the percentiles of the hypothesized probability density. The kernel used is the Gaussian density with the assumed variance based on the observed data. The kernel density curve determines the smoothness of the resulting estimates. The bottom part of the chart is the box plot of the data and the 95% confidence intervals.

The Q–Q Plots for Performance chart shows how the data is distributed. The horizontal axis is the quantile value, and the vertical axis represents the performance ratings. The solid diagonal line represents data distribution following a perfect normal Gaussian pattern. The empty circle represents the observed ratings which follow closely with the solid line, indicating its normal distribution characteristics.

6.5 CORRELATION DETERMINATION – PROC CORR

Correlation describes the strength of the relationship between variables. Some relationships are strong and therefore have a strong correlation. Some relationships between variables are weak; hence, we say there is a weak correlation. For relationships that do not exist between variables, we state there is no correlation. In statistical terms, we define the strength of relationship between two variables as the correlation coefficient. The correlation coefficient ranges from +1 to −1 and is often presented as the "r" value.

The closer the r to either 1 or −1, the stronger is the correlation between the two variables evaluated. When r = 1 or −1, the correlation is said to be "perfect." When r = 0, the correlation does not exist at all. When the r-value is positive, the correlation is positive, meaning that when one variable increases, the other one has the tendency to increase. When r-value is negative, the correlation is negative, meaning that when one variable increases, the other one tends to decrease.

There are many different r-value types based on different computation methods and input data characteristics. The most common ones include the Pearson correlation coefficient, the Spearman correlation coefficient, and the Kendall correlation coefficient.

Be careful with correlation interpretations. You must remember that correlation does not mean causation. For example, when r = +1 and both variables move together, this does not necessary mean that the increase in one variable is caused by the corresponding variable in your correlation study.

Data: transport.xlsx

The transport.xlsx data has six variables: year (1961–2015 (55 years)), annual total gasoline fuel consumption (million gallons), number of registered vehicles (million vehicles), population (million people), licensed drivers (million drivers), and public roadway length (miles).

Year	Fuel	RoadLength	Population	Drivers	Vehicles
1961	60006	3573046	180	87	74
1962	62204	3599581	183	89	76
1963	64968	3620457	186	91	79
1964	68318	3644069	188	94	83
1965	71592	3689666	191	95	86
1966	75475	3697950	194	99	90
1967	78621	3704914	196	101	94
1968	83746	3684085	197	103	97
1969	88935	3710299	199	105	101
1970	92967	3730082	201	108	105
1971	98150	3758942	204	112	108

Task: Determine whether there is any significant correlation between the five variables listed in the transport.xlsx data.

Analysis: SAS's PROC CORR is the right tool to perform the test. We will do two separate tests to illustrate how the PROC CORR works and its results.

Case A: Correlation between population and drivers. Intuition tells us there should be a strong correlation between them – more people means more licensed drivers.

Case A: SAS Codes (E65a)

```
TITLE "PROC Corr-Transport Data";
LIBNAME myt 'c:\sas_exercise';

PROC IMPORT OUT= WORK.transport
            DATAFILE= "c:\sas_exercise\transport.xlsx"
            DBMS=EXCEL REPLACE;
      RANGE="Sheet1$";
      GETNAMES=YES;
      MIXED=NO;
      SCANTEXT=YES;
      USEDATE=YES;
      SCANTIME=YES;
RUN;

PROC CORR
DATA=transport ;
VAR Population
Drivers;
RUN;
```

Explanations for codes from the TITLE statement until the PROC CORR statement can be obtained from Section 3.5.1 Importing an Excel File.

PROC CORR DATA=transport invokes the CORR procedure for the dataset named transport.sas7bdat in the WORK library.

VAR declares that variables to be analyzed are Population and Drivers. It will determine the correlation strength between Population and Drivers.

RUN instructs SAS to execute the PROC CORR procedure.

Case A Result (E65a):

1 ***PROC Corr-Transport Data***

2 ***The CORR Procedure***

3

2 Variables:	Population Drivers

4

Simple Statistics							
5	6	7	8	9	10	11	12
Variable	**N**	**Mean**	**Std Dev**	**Sum**	**Minimum**	**Maximum**	**Label**
Population	55	246.60000	41.29057	13563	180.00000	319.00000	Population
Drivers	55	156.98182	39.60008	8634	87.00000	214.00000	Drivers

13

Pearson Correlation Coefficients, N = 55 Prob > |r| under H0: Rho=0		
	Population	**Drivers**
Population Population	1.00000	15 0.98665 <.0001
Drivers Drivers	14 0.98665 <.0001	1.00000

1 – it is the title specified.

2 – it is the default subtitle for the PROC CORR procedure.

3 – variables specified: Population and Drivers.

4 – a list of common statistics measures.

5 – the variable column lists each of the two variables.

6 – the total number of observations for each of the variables.

7 – it is the arithmetic average $(\Sigma (x_i))/N$.

8 – it is the standard deviation and equals to the square root of the variance. It measures central tendency.

9 – it is the sum of all observations for each of the variables.

10 – the smallest observed value.

11 – the largest observed value.

12 – SAS can display a variable with a different name by using the Label command. Here the names are the name as under Variable meaning no label command is used.

13 – this table contains the computed Pearson correlation coefficients between two parameters. The coefficient measures the strength of the correlation between the variable Population and Drivers.

Prob>|r| under Ho: Rho = 0 – it refers to the p value. Under the assumption that the null hypothesis is true (the correlation coefficient equals to 0: no correlation at all), the probability to observe the computed coefficients as they are or anything larger than that due to chance is the p value.

14 and 15 – the Pearson correlation coefficients between Population and Drivers are 0.98665 based on 55 (N = 55) observations with a p value of less than 0.0001. The null hypothesis is rejected, and it is concluded that the Population is statistically correlated with Drivers in a significant way.

Case B: Correlation between Fuel, Driver, and Vehicles

Case B: SAS Codes (E65b) - with Plot Options Specified

```
TITLE "PROC Corr-Transport Data";
LIBNAME myt 'c:\sas_exercise';

PROC IMPORT OUT= WORK.transport
            DATAFILE= "c:\sas_exercise\transport.xlsx"
            DBMS=EXCEL REPLACE;
     RANGE="Sheet1$";
     GETNAMES=YES;
     MIXED=NO;
     SCANTEXT=YES;
     USEDATE=YES;
     SCANTIME=YES;
RUN;

PROC CORR DATA=transport
plots=matrix(histogram);
VAR Fuel Drivers
Vehicles;
RUN;
```

Explanations for codes from the TITLE statement until the PROC CORR statement can be obtained from Section 3.5.1 Importing an Excel File.

PROC CORR DATA=transport invokes the CORR procedure for a dataset named transport in the WORK library. Plots=matrix (histogram) is an option that instructs SAS to plot out a series of histograms for variables specified.

VAR declares that the variables are Fuel, Drivers, and Vehicles.

RUN instructs SAS to execute the PROC CORR procedure.

Case B Result (E65b):

PROC Corr-Transport Data 07:08 Thursday,

The CORR Procedure

3 Variables:	Fuel Drivers Vehicles

Simple Statistics							
Variable	N	Mean	Std Dev	Sum	Minimum	Maximum	Label
Fuel	55	130404	35155	7172200	60006	178915	Fuel
Drivers	55	156.98182	39.60008	8634	87.00000	214.00000	Drivers
Vehicles	55	172.81818	56.21918	9505	74.00000	252.00000	Vehicles

Pearson Correlation Coefficients, N = 55 Prob > |r| under H0: Rho=0			
	Fuel	Drivers	Vehicles
Fuel Fuel	1.00000	0.98096 <.0001	0.98420 <.0001
Drivers Drivers	0.98096 <.0001	1.00000	0.99849 <.0001
Vehicles Vehicles	0.98420 <.0001	0.99849 <.0001	1.00000

The interpretation of the E65b result is the same as E65a. Additional information that E65b has are the histograms plotted out between two of the three variables. Each of these additional histograms provides a visual illustration of how the 65 data pairs trend.

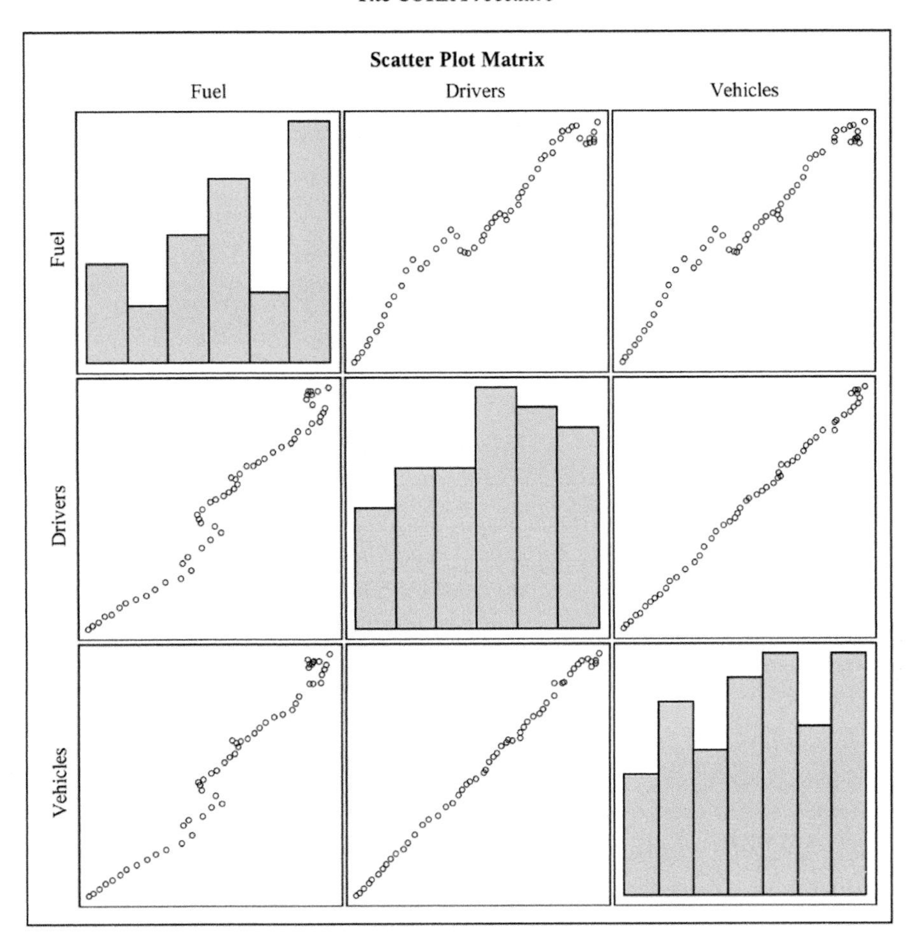

6.6 REGRESSION ANALYSIS – PROC REG

Regression analysis is a statistical modeling tool used for estimating a relationship between a dependent variable and one or more independent variables. The coefficients of the independent variables in the regression equation indicate how the dependable variable changes when the corresponding independent variable changes by one unit.

The p value for each coefficient tells you whether the coefficient is statistically and significantly different from zero. If your p value is less than your α significance level, you can reject the null hypothesis. In other words, the coefficient being not 0 is not the result of random chance.

As with correlation analysis, regression analysis can also provide estimates of the correlation between one phenomenon and another. Just as how correlation does not mean causation, regression analysis also does not mean causation. However, unlike correlation analysis, which is a simple bivariant analysis, regression does appear to possess some causation connotation.

For regression analysis, you put a lot of effort into determining what measure should be the dependent variable and what measures should be the independent variables – parameters that may have a causal effect on the dependent variable. This is where the subject matter expertise and critical thinking should come into play.

In terms of computational approaches, the most common method of regression analysis is the ordinary least square (OLS).

SAS's PROC REG specification is straightforward, as illustrated below. The PROC REG uses the OLS to create a regression model. If you want to fit a particular model to your data, you need to specify your model(s) by the MODEL option. The MODEL option specifies both the dependent and explanatory variables.

The PROC REG can't handle discrete categorical classification variables. To deal with discrete categorical classification variables, you need to assign numerical values to different categories (parameterize the observations). Also, the PROC REG can't handle interaction effects among independent variables (Section 6.7 has a brief discussion on what interaction effect is).

Example: 6.6

Case A: All dependent variables are quantitative measures

Data: transport.xlsx

The transport.xlsx data has six variables: year (1961–2015 (55 years)), annual total fuel consumption (million gallons), number of registered vehicles (million vehicles), population (million people), licensed drivers (million drivers), and public roadway length (miles). The data are illustrated below.

Year	Fuel	RoadLength	Population	Drivers	Vehicles
1961	60006	3573046	180	87	74
1962	62204	3599581	183	89	76
1963	64968	3620457	186	91	79
1964	68318	3644069	188	94	83
1965	71592	3689666	191	95	86
1966	75475	3697950	194	99	90
1967	78621	3704914	196	101	94
1968	83746	3684085	197	103	97
1969	88935	3710299	199	105	101
1970	92967	3730082	201	108	105
1971	98150	3758942	204	112	108

Task: Assess how the number of vehicles, population, and length of public roads affects fuel consumption.

Analysis: In the analysis here, two different models are specified to illustrate how to specify a model. The first model is fuel = f (vehicles only), and the second is fuel = f (vehicles population roadlength).

SAS Codes (E66a)

```
TITLE "PROC Reg-Transport Data";
LIBNAME myt 'c:\sas_exercise';

PROC IMPORT OUT= WORK.transport
            DATAFILE= "c:\sas_exercise\transport.xlsx"
            DBMS=EXCEL REPLACE;
     RANGE="Sheet1$";
     GETNAMES=YES;
     MIXED=NO;
     SCANTEXT=YES;
     USEDATE=YES;
     SCANTIME=YES;
RUN;

PROC REG data =transport  RSQUARE plots=none;
MODEL  fuel=vehicles;
MODEL fuel = vehicles population roadlength;
RUN;
```

Explanations for codes from the TITLE statement until the PROC REG statement can be obtained from Section 3.5.1 Importing an Excel File.

PROC REG invokes the Reg procedure for a dataset named transport.sas7bdat in the WORK library. The RSQUARE is an option that instructs SAS to compute the R-squared value.

MODEL declares that fuel is a dependent variable and vehicles is an independent variable.

The next MODEL (model 2) statement declares that fuel is a dependent variable, while vehicles, population, and roadlength are the independent variables.

RUN instructs SAS to execute the PROC REG procedure.

Result (E66a):

1 ***PROC Reg-Transport Data***

2 ***The REG Procedure***

3 ***Model: MODEL1***

4 ***Dependent Variable: Fuel Fuel***

5	Number of Observations Read	55
6	Number of Observations Used	55

7 Analysis of Variance					
Source 8	9 DF	10 Sum of Squares	11 Mean Square	12 F Value	13 Pr > F
Model 14	1	64645690298	64645690298	1637.50	<.0001
Error 15	53	2092346279	39478232		
Corrected Total 16	54	66738036577			

			18		
	Root MSE 17	6283.17051	R-Square	0.9686	
19	Dependent Mean	130404	Adj R-Sq	0.9681	20
21	Coeff Var	4.81825			

22 Parameter Estimates						
23 Variable	24 Label	25 DF	26 Parameter Estimate	27 Standard Error	28 t Value	29 Pr > |t|
Intercept	30 Intercept	1	24044	2761.54446	8.71	<.0001
Vehicles	31 Vehicles	1	615.44382	15.20889	40.47	<.0001

1 – it is the title specified.

2 – it is the default subtitle for the PROC REG procedure.

3 – it refers to the first model specified: Model Fuel = Vehicles.

4 – it identifies the dependent variable, which is Fuel.

5 and 6 – 55 are the number of observations (rows) read and used in the analysis, respectively. The Number of Observations Used may be fewer than the Number of Observations Read if missing values are encountered.

7 – this table provides information on the result of variance analysis.

8 – this column lists the sources of the variance.

9 – degree of freedom.

10 – the Sum of Squares has 3 parts. The 1st row is the Model SS. The second row is Error SS. And the last row is the Corrected Total SS. Model SS = SUM $(Y_{model} - \bar{Y})^2$*. Error SS = SUM* $(Y - Y_{model})^2$*. And Corrected Total SS = SUM* $(Y - \bar{Y})^2$*.*

11 – it is the result of dividing the Sum of Squares by the DF. It has two parts: the Mean Square for the Model and the Mean Square for the Error (Residual).

12 – it is the F Statistic and computed by dividing the Model SS by the Error SS.

13 – it refers to the p value as related to the question of how reliable the independent variable (here it is the number of vehicles) predicts the dependent variable (here it is the amount of fuel consumed). In this example, the p value is < 0.0001. The hypothesis is rejected, and it is concluded that the number of vehicles does predict the amount of fuel consumed.

14 – it refers to the Model 1 specified. Relevant F statistics are provided in this row for the model component.

15 – the amount of variation which can't be explained by the specified model.

16 – the combined amount from model and error components.

17 – it is the standard deviation of the error term. It equals to the square root of the mean square for error.

18 – it is the percentage of variance associated with the dependent variable that is predicted from the independent variables. In this example, it is 96.86%. It reflects the explanation power of the entire model.

19 – it is the mean of the dependent variable. In this example, it is fuel.

20 – the R-Square of a model will get bigger as more independent variables are added even when such additional term has no real effect (however, simply chance also helps). SAS computes an Adjusted R-Squared with the formula of $1 - ((1 - Rsq)((N - 1)/(N - K - 1))$, where N is the number of observed values and K is the number of independent variables. The Adj R – Sq value is an attempt to provide a more truthful estimation on a model's explanation power.

21 – it refers to the coefficient of variation, and it measures data variation. It is based on the formula of Coeff Var = (root mean square)/(mean of the dependent variable) × 100. The Coeff Var is unitless. In this example, it is $6283.17051/130404 \times 100 = 4.81825$.

22 – it lists parameters estimated as related to the model specified. In this case, model 1 specified is fuel = vehicle. The regression equation takes the form of $y = a + bx$.

23 – lists the model variables. Here, the model has its Intercept and independent variable Vehicles.

24 – SAS can display a variable with a different name by using the Label command. Here, the names are the same. No label command is used.

25 – degree of freedom

26 – it lists the parameters SAS computed, including the intercept and the coefficients for the independent variables. In this case, the actual model can be expressed as Fuel = 24044 + 615.44382 × (Vehicles).

27 – it lists the standard error associated with each coefficient. It is needed for t value and confidence interval computation usages.

28 – it is the t value gained by dividing the Parameter Estimate with the Standard Error.

29 – it lists the two-tailed p value for the null hypothesis (coefficient and parameter = 0) testing. In this example, the p values are less than 0.0001 for both the Intercept and Vehicles. Both null hypotheses are rejected.

30 and 31 – information related to model 1 slope and intercepts is provided.

Model: Model2 – refers to the 2nd model specified: Model Fuel = Vehicles Population Roadlength

The interpretation of the output is similar to Model 1. The eventual Model 2 is:

Fuel = −157929 + 642.16379 × Vehicles − 239.20593 × Population + 0.06078 × RoadLength

The REG Procedure
Model: MODEL2
Dependent Variable: Fuel Fuel

Number of Observations Read	55
Number of Observations Used	55

Analysis of Variance					
Source	DF	Sum of Squares	Mean Square	F Value	Pr > F
Model	3	64997490734	21665830245	634.83	<.0001
Error	51	1740545843	34128350		
Corrected Total	54	66738036577			

Root MSE	5841.94744	R-Square	0.9739
Dependent Mean	130404	Adj R-Sq	0.9724
Coeff Var	4.47990		

Parameter Estimates						
Variable	Label	DF	Parameter Estimate	Standard Error	t Value	Pr > |t|
Intercept	Intercept	1	-157929	64518	-2.45	0.0179
Vehicles	Vehicles	1	642.16376	87.03513	7.38	<.0001
Population	Population	1	-239.20593	124.87029	-1.92	0.0610
RoadLength	RoadLength	1	0.06078	0.01967	3.09	0.0032

Case B: One of the Dependent Variables Is Qualitative in Nature

Data: trip. xlsx
The trip.xlsx data contains the survey result of 100 families. It has data of trip, family size (fs), homeownership status (hos), and income level (income). The trip variable is the average number of trips a person makes in a day. Family size (fs) is the number of individuals a family has. Homeownership (hos) is a binary variable with either a Yes or No response depending on whether a family owns their living quarters. Income is a categorical variable and has the designation of low, medium, or high.

This dataset is not balanced due to varied **fs** observations.

trip	income	hos	fs
5.2	low	yes	3
6.8	medium	yes	2
6.8	high	yes	2
4.9	low	no	3
6.4	medium	no	3
6.5	high	no	3
5.1	low	yes	3
6.6	medium	yes	4
6.6	high	yes	3
4.8	low	no	4
6.2	medium	no	3
6.3	high	no	3
5.2	low	yes	1
6.4	medium	yes	3
6.4	high	yes	2
4.6	low	no	4
6.1	medium	no	3
6.1	high	no	2
4.8	low	yes	1
6.2	medium	yes	4
6.2	high	yes	5
4.5	low	no	6
5.8	medium	no	4
5.9	high	no	1
4.7	low	yes	3
6.2	medium	yes	2
6.2	high	yes	4
4.3	low	no	2
5.7	medium	no	2
5.8	high	no	1
4.6	low	yes	4
5.8	medium	yes	5
5.8	high	yes	3
5.2	low	no	3
6.9	medium	no	2
6.2	high	no	1
5.5	low	yes	1
7.1	medium	yes	1
7.1	high	yes	2
6.4	low	no	3
7.3	medium	no	1
7.3	high	no	1
5.4	low	yes	2
7.1	medium	yes	3
7.1	high	yes	2
5.1	low	no	3
6.7	medium	no	3
6.8	high	no	2
5.3	low	yes	4
6.9	medium	yes	3
6.9	high	yes	3
6.2	low	no	3
8.1	medium	no	4
8.2	high	no	3
6.4	low	yes	4
7.3	medium	yes	3
7.3	high	yes	1
6.1	low	no	3
7.9	medium	no	4
7.1	high	no	2

Task:
Determine how the number of vehicle trips (trips) per person is affected by

- Family size (fs)
- Homeownership status (hos)
- Income (income).

Analysis: We will need to parameterize the "income" and "hos" from a categorical rating to numerical assignments.

SAS Codes (E66b)

```
TITLE 'PROC Reg-Trip Data';
LIBNAME myt 'c:\sas_exercise';
PROC IMPORT OUT= WORK.trip
            DATAFILE= "c:\sas_
exercise\trip.xlsx"
            DBMS=EXCEL REPLACE;
     RANGE="trip$";
     GETNAMES=YES;
     MIXED=NO;
     SCANTEXT=YES;
     USEDATE=YES;
     SCANTIME=YES;
RUN;
DATA myt.trip2;
SET trip;
IF hos='yes' THEN hos2=1;
ELSE IF hos='no' THEN hos2=2;
IF income='high' THEN income2=1;
ELSE IF income='medium' THEN
income2=2;
ELSE IF income='low' THEN
income2=3;
RUN;
PROC REG data =myt.trip2
RSQUARE plots=none;
MODEL trip=fs;
MODEL trip=hos2;
MODEL trip=income2;
MODEL trip=fs hos2 income2;
RUN;
```

Explanations for codes from the TITLE statement until the DATA myt.trip2; statement can be obtained from Section 3.5.1 Importing an Excel File.

DATA creates a new SAS datafile named trip2 and stores it in the myt library.

SET tells SAS that the trip2 dataset is to be based on the trip data in the WORK library.

IF hos="yes" THEN hos2=1 is a conditional statement creating a new variable called hos2.

ELSE IF hos="no" THEN hos2=2 is the other conditional statement.

IF income="high" THEN income2=1 creates a new variable called income2.

Statements of ELSE IF … THEN fulfill alternative income conditions.

All the above IF THEN and ELSE IF THEN statements create two numerical variables hos2 and income2 based on two categorical variables of hos and income.

Run kicks off the data step.

PROC REG invokes the REG function for the dataset named trip2.sas7bdat in the myt library. The RSQUARE is an option that instructs SAS to compute the R-squared value.

MODEL trip=fs declares that trip is a dependent variable and fs is an independent variable.

MODEL trip=hos2 (Model 2) declares that trip is a dependent variable and hos2 is an independent variable.

MODEL trip=income2 (Model 3) declares that trip is a dependent variable and income2 is an independent variable.

MODEL trip=fs hos2 income2 (Model 4) declares that trip is a dependent variable and fs, hos2 and income2 are the independent variables.

RUN kicks off the PROC REG procedure.

Results from All Models (E66b):

Interpretations for all the results follow the same definitions used in E66a.

Result from Model 1 (Model trip = fs)

Model: MODEL1
Dependent Variable: trip trip

Number of Observations Read	60
Number of Observations Used	60

Analysis of Variance					
Source	DF	Sum of Squares	Mean Square	F Value	Pr > F
Model	1	1.53799	1.53799	1.82	0.1828
Error	58	49.07934	0.84620		
Corrected Total	59	50.61733			

Root MSE	0.91989	R-Square	0.0304
Dependent Mean	6.17333	Adj R-Sq	0.0137
Coeff Var	14.90101		

Parameter Estimates						
Variable	Label	DF	Parameter Estimate	Standard Error	t Value	Pr > |t|
Intercept	Intercept	1	6.56136	0.31136	21.07	<.0001
fs	fs	1	-0.14110	0.10466	-1.35	0.1828

Results from MODEL trip = hos2 (model 2) and MODEL trip = income2 (Model 3)

Results from Models 2 and 3 have the same format as Model 1. They are omitted.

Results from Models **MODEL trip = fs hos2 income** (Model 4) are listed below. Their interpretation is the same as for Model 1.

Model: MODEL4
Dependent Variable: trip trip

Number of Observations Read	60
Number of Observations Used	60

Analysis of Variance					
Source	DF	Sum of Squares	Mean Square	F Value	Pr > F
Model	3	20.03842	6.67947	12.23	<.0001
Error	56	30.57891	0.54605		
Corrected Total	59	50.61733			

Root MSE	0.73895	R-Square	0.3959
Dependent Mean	6.17333	Adj R-Sq	0.3635
Coeff Var	11.97009		

Parameter Estimates						
Variable	Label	DF	Parameter Estimate	Standard Error	t Value	Pr > |t|
Intercept	Intercept	1	7.59703	0.42297	17.96	<.0001
fs	fs	1	-0.01366	0.08689	-0.16	0.8756
hos2		1	0.01288	0.19082	0.07	0.9464
income2		1	-0.70272	0.12073	-5.82	<.0001

6.7 PERFORMING THE GENERAL LINEAR MODELING – PROC GLM

GLM refers to linear regression modeling of a continuous dependent variable (result or response) to a family of independent variables (explanatory parameters or predictors) presented as either continuous or categorical formats.

Two of the key applications of regression analysis are causation and prediction. For causation determination, no statistical work, including regression alone, can conclude causation. However, the variance analysis, as carried out in the GLM procedure, does help to rule out the alternative hypothesis (for example, the null hypothesis Ho is $\beta 1 = 0$. And the p value for var1 is 0.0056. We reject the null hypothesis that the coefficient for var1 is zero, and we accept the alternative hypothesis that the coefficient is somewhat different from 0).

Assessing the significance of the coefficient of each independent variable during a causation analysis is the typical focus. During the process, you may discover that some of the explanatory variables specified in a model trend the same way, resulting in the explanatory power of the specified model being much less than the sum of each individual variable's explanatory power. This phenomenon, known as multicollinearity, indicates the presence of a strong correlation among independent variables. In a scenario like this, appropriate treatments should be carried out to minimize such impact.

For the predictive analysis, the focus is on the coefficient of determination, also known as R-squared (R^2). The R^2 indicates how much of the response is explained by the specified model. Obviously, the larger the R^2, the more desiring with all regression analyses.

SAS's PROC GLM procedure covers a wide range of analytical capacities, including:

- Simple regression (linear: $y = f(x)$)
- Multiple regression (linear: $y = f(x_1, x_2, x_3...)$)
- Polynomial regression (linear or nonlinear $y = f(x_1, x_1^n, x_1^m, x_2, x_2^k, x_2^L ...)$)
- Analysis of variance (ANOVA)
- Analysis of covariance
- Others.

The PROC GLM handles categorical variables directly through the statement of CLASS. The PROC GLM can also handle interaction effects among the independent variables.

An interaction effect refers to how the dependent variable's behavior, in response to one independent variable, is dependent upon the intensity of another independent variable. For example, the number of vacation trips a family makes in a year is impacted by the number of vehicles a low-income family has. However, as family income level changes from "low" to "medium" and then to "high," the effect from the number of vehicles a family owns diminishes. Consequently, we can't simply that state vehicle ownership affects vacation trips.

You should always start to examine the interaction effects among variables first. If any interaction effect is significant, you should first attempt to explain the interaction effect. One last aspect of the interaction effect is that it is extremely challenging to explain a phenomenon with more than three dimensions. Your specification for any interaction effect should be limited to no more than three levels (var1*var2*var3).

Example: 6.7

Data: trip.xlsx
The trip.xlsx has the following variables: trip, family size (fs), homeownership status (hos), and income level (income). It has the survey results of 100 families. The trip is the average number of trips a person makes in a day. Family size (fs) is the number of people a family has. Homeownership (hos) is a Yes or No choice on whether a family owns his or her living quarters. Income is a qualitative measure and has the destination of low, medium, or high. See below for the full set of data.

The data is not a balanced dataset due to varied **fs** observations.

trip	income	hos	fs
5.2	low	yes	3
6.8	medium	yes	2
6.8	high	yes	2
4.9	low	no	3
6.4	medium	no	3
6.5	high	no	3
5.1	low	yes	3
6.6	medium	yes	4
6.6	high	yes	3
4.8	low	no	4
6.2	medium	no	3
6.3	high	no	3
5.2	low	yes	1
6.4	medium	yes	3
6.4	high	yes	2
4.6	low	no	4
6.1	medium	no	3
6.1	high	no	2
4.8	low	yes	1
6.2	medium	yes	4
6.2	high	yes	5
4.5	low	no	6
5.8	medium	no	4
5.9	high	no	1
4.7	low	yes	3
6.2	medium	yes	2
6.2	high	yes	4
4.3	low	no	2
5.7	medium	no	2
5.8	high	no	1
4.6	low	yes	4
5.8	medium	yes	5
5.8	high	yes	3
5.2	low	no	3
6.9	medium	no	2
6.2	high	no	1
5.5	low	yes	1
7.1	medium	yes	1
7.1	high	yes	2
6.4	low	no	3
7.3	medium	no	1
7.3	high	no	1
5.4	low	yes	2
7.1	medium	yes	3
7.1	high	yes	2
5.1	low	no	3
6.7	medium	no	3
6.8	high	no	2
5.3	low	yes	4
6.9	medium	yes	3
6.9	high	yes	3
6.2	low	no	3
8.1	medium	no	4
8.2	high	no	3
6.4	low	yes	4
7.3	medium	yes	3
7.3	high	yes	1
6.1	low	no	3
7.9	medium	no	4
7.1	high	no	2

Task: Determine how family income level, homeownership status, and family size affect the number of trips a person makes in a day.

SAS Codes (E67)

```
TITLE "PROC GLM - Trip Data";
LIBNAME myg 'c:\sas_exercise';

PROC IMPORT OUT= WORK.trip
            DATAFILE= "c:\sas_
exercise\trip.xlsx"
            DBMS=EXCEL REPLACE;
     RANGE="trip$";
     GETNAMES=YES;
     MIXED=NO;
     SCANTEXT=YES;
     USEDATE=YES;
     SCANTIME=YES;

DATA tripG;
SET trip;

PROC GLM data =tripG  plots=none;
CLASS hos income;
MODEL trip= fs  hos income  fs*hos
fs*income
hos*income fs*hos*income;
RUN;
```

Explanations for codes from the TITLE statement until the DATA tripG; statement can be obtained from Section 3.5.1 Importing an Excel File.

PROC GLM invokes the GLM procedure for a dataset named tripG in the WORK library.

CLASS declares that hos and income are the categorical variables.

*MODEL specifies that trip is the dependent variable, and it may be affected by variables fs, hos, income, and interaction effects, including fs*hos, fs*income, hos*income, and fs*hos*income.*

RUN instructs SAS to execute the PROC GLM procedure.

Result (E67):

***1** PROC GLM - Trip Data*

***2** The GLM Procedure*

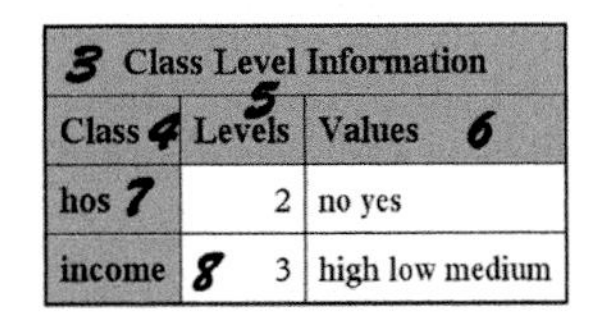

3 Class Level Information		
Class 4	Levels 5	Values 6
hos 7	2	no yes
income 8	3	high low medium

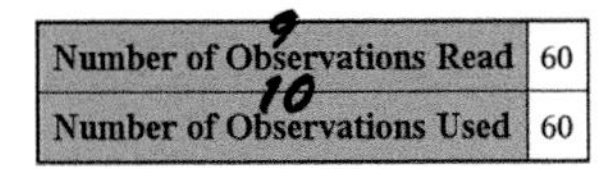

Number of Observations Read 9	60
Number of Observations Used 10	60

1 – it is the title specified.

2 – the default subtitle for the PROC GLM procedure.

3 – it provides information on the qualitative variables.

4 – the Class has two variables: hos and income.

5 – the Levels lists the number of levels each qualitative variable has. In this example, the hos has two levels (yes and no) and the income has three levels (low, medium, and high).

6 – the Values column provides what each level has. hos have yes and no. And income has high, low, and medium.

7 and 8 – variables hos and income are shown.

11 PROC GLM - Trip Data

12 The GLM Procedure

Dependent Variable: trip trip 13

14 Source	15 DF	16 Sum of Squares	17 Mean Square	18 F Value	19 Pr > F
Model 20	11	30.81981156	2.80180105	6.79	<.0001
Error 21	48	19.79752178	0.41244837		
Corrected Total 22	59	50.61733333			

23 R-Square	24 Coeff Var	25 Root MSE	26 trip Mean
0.608879	10.40315	0.642221	6.173333

27 Source	28 DF	29 Type I SS	30 Mean Square	31 F Value	32 Pr > F
fs 33	1	1.53799353	1.53799353	3.73	0.0594
hos 34	1	0.00111739	0.00111739	0.00	0.9587
income 35	2	26.26063139	13.13031569	31.84	<.0001
fs*hos 36	1	0.66072923	0.66072923	1.60	0.2117
fs*income 37	2	0.00558993	0.00279496	0.01	0.9932
hos*income 38	2	0.01155876	0.00577938	0.01	0.9861
fs*hos*income 39	2	2.34219134	1.17109567	2.84	0.0683

40 Source	41 DF	42 Type III SS	43 Mean Square	44 F Value	45 Pr > F
fs 46	1	0.05764323	0.05764323	0.14	0.7102
hos 47	1	0.54870209	0.54870209	1.33	0.2545
income 48	2	2.36909297	1.18454648	2.87	0.0664
fs*hos 49	1	0.95345042	0.95345042	2.31	0.1350
fs*income 50	2	0.12929851	0.06464926	0.16	0.8554
hos*income 51	2	1.87115566	0.93557783	2.27	0.1145
fs*hos*income 52	2	2.34219134	1.17109567	2.84	0.0683

9 and 10 – 60 are the number of observations (rows) read and used in the analysis, respectively. The Number of Observations Used may be fewer than the Number of Observations Read if missing values are encountered.

11 – it is the title specified (this is the second page of the output).

12 – it is the default subtitle of the PROC GLM procedure.

13 – it identifies that the dependent variable is trip.

14 – this column outlines the model components.

15 – degree of freedom.

16 – it refers to the various Sum of Squares. And it has three parts. The first row is the Model SS. The second row is Error SS. And the last row is the Corrected Total SS. Model SS = SUM $(Y_{model} - \bar{Y})^2$. Error SS = SUM$(Y - Y_{model})^2$. And the Corrected Total SS = SUM $(Y - \bar{Y})^2$.

17 – it is the result of dividing the Sum of Squares by the DF. It has two parts: Mean Square for the Model and Mean Square for the Error (Residual).

18 – it is the F Statistics and computed by dividing the Model SS by the Error SS.

19 – it refers to the p value as related to the question of how reliable the independent variable predicts the dependent variables. In this example, the p value is < 0.0001. The hypothesis is rejected. And it is concluded that the independent variables do predict the dependent variable.

20 – the amount of variation can be explained by the specified model.

21 – the amount of variation which can't be explained by the specified model.

22 – the combined amount from both the Model and Error components.

23 – it is the percentage of dependent variable variance that is predicted from the independent variables. In this example, it is 60.8879%. It reflects the explanation power of the entire model.

24 – it refers to the coefficient of variation measuring data variation. It is based on the formula of Coeff Var = (root mean square)/ (mean of the dependent variable) × 100. The Coeff Var is unitless. In this example, it is 0.642221/6.1733333 × 100 = 10.40315.

25 – it refers to the standard deviation of the error term. It equals to the square root of the Mean Square for Error.

26 – it is the mean of the dependent variable trip.

27 – this column lists all the independent variables and their interaction terms (model components).

28 – degree of Freedom for each Source.

29 – by default, the PROC GLM procedure generates both Type I SS and Type III SS. The Type I SS is also known as the sequential SS. This is due to the fact that as each additional Source (independent variable) is added to the model, the effect from this newly added Source is adjusted by all the Sources already presented in the model, resulting in a decrease in the Error SS. The order of the Source appearance affects the amount of SS associated with the Source. The Source added later will have less variance to account for because more are already accounted for by previous Sources. For a set of balanced data, Type I SS is the same as Type III SS. Type I SS is more appropriate for balanced data given that the frequency of observations are weighted (for balanced, the frequency is the same for each level of observation). For this example, given the data is not balanced, you can ignore the Type I SS and go to Type III SS directly.

Type III SS – compared with the Type I SS, each Type III SS is adjusted for all other sources in the model, regardless of order. Type III SS treats each Source as if it were the last term in a fully specified model. The frequency of any observations (missing value problem) for a given level does not affect the magnitude of the Type III SS.

30 – it is calculated based on the formula of SS/DF.

31 – it is the F statistics gained by dividing the Source Mean Square by the Error Mean Square for each Source (independent variable).

32 – it lists the p value for the null hypothesis (coefficient = 0) testing.

33, 34, 35, 36, 37, 38, and 39 – F statistics computed for each model component in a sequential order are provided here. Both interaction and main effects are tested. When any interaction effect is specified in a model, the explanation of the result should always start with the interaction effect. Main effect should only be discussed when the interaction effect is not significant.

In this example, we should start with the three-way interaction effect where its p value is 0.0683. This value is legitimate to be considered in rejecting the null hypothesis that there is no significant three-way interaction effect. It means the impact from fs, hos, and income to trip are not consistent depending on the levels of each independent variable.

Whenever a significant interaction effect exists, the interpretation of the main effect as explained below should be stopped. Further analysis of such interaction effect should be carried out for each level of Source (e.g., for each income level (income), conduct a separate analysis, and draw conclusions for that level of income group).

The explanation below should not be carried out given the three-way interaction effect is significant. However, for illustration purpose, explanations are offered below by pretending that the three-way interaction effect was not significant.

40 to 45 – refer to 27–32 and pay particular attention to 29 on Type III SS.

49, 50, and 51 – the p values for all the two-way interactions are large, and it is failed to reject the hypothesis that the effects between each of the two variables impact each other in a statistical significant manner. There are no significant two-way interaction effects.
46 – the p value for fs is 0.0594. The hypothesis is rejected. The fs has a significant effect on trip. An effect from the fs is called the main effect.
47 – the p value for hos is 0.9587. The hypothesis is failed to be rejected. The hos has no significant effect on trip. An effect from the hos is called the main effect.
48 – the p value for income is less than 0.0001. the hypothesis is rejected. The income has a significant effect on trip. The income factor is one of the main effects on the trip rate.

6.8 ANALYSIS OF VARIANCE – PROC ANOVA

ANOVA stands for the analysis of variance. It is an extension of the Student t-test and the z-test. However, unlike the t-test, ANOVA enables you to perform mean testing on more than two different samples simultaneously. The analysis splits the total variation represented by the independent variable (response) into a systematic component due to the specified independent variables (explanatory) and a random error component. It answers the question of how independent variables affect the dependent variable.

You will notice the terms "one-way ANOVA" and "two-way ANOVA." One-way ANOVA refers to the case where there is only one independent variable. On the other hand, two-way ANOVA is where there are two independent variables specified in a model. The two-way ANOVA can examine the interaction effect between two independent variables. Keep in mind that ANOVA can be used in situations with more than two independent variables. But rarely it is used with more than 3 independent variables as results are complex to interpret.

ANOVA is also commonly known as the F-test, after the method's developer, Ronald Fischer. The computation produces an F-value (also known as F-statistics) called ANOVA coefficient based on the formula of:

$F = MST/MSE$

where

MST: the mean sum of squares due to independent variables
MSE: the mean sum of squares due to random errors.

Using the F-value and degrees of freedom, a p value is generated for each of your variables. You draw conclusions based on the p value with regard to accepting or rejecting your null hypothesis (the null hypothesis is that there is no difference).

SAS's PROC ANOVA is designed for balanced data based on randomized experimental designs (for non-balanced data, the PROC GLM should be used). Below offers a simple illustration of the concept of balanced and none balanced data.

TABLE 6.1

Number of Families Participated in the Trip Study – Balanced Data

		Family Income Level		
		Low	**Medium**	**High**
Homeownership status	Yes	10 Families	10 Families	10 Families
	No	10 Families	10 Families	10 Families

TABLE 6.2

Number of Families Participated in the Trip Study – Unbalanced Data

		Family Income Level		
		Low	**Medium**	**High**
Homeownership status	Yes	6 Families	10 Families	9 Families
	No	10 Families	7 Families	10 Families

You are planning to analyze how homeownership status and income affect the number of daily trips a person takes. You recruit families where you can record such trip information (dependent variable) with the following attributes: a) family income level: low, medium, and high, and b) homeownership status: yes or no. Here the family income level and homeownership status are called classification variables (categorical), which are independent variables.

A balanced dataset is a set of data where all "levels" of the data are equally represented. An unbalanced dataset is the opposite. For the dataset to be balanced, each family must have an observation for each variable group.

For example, if your data is balanced, your daily trip study will include an equal number of families participating in each scenario (the combination of income level and homeownership status) as illustrated in Table 6.1. You should always strive for balanced data from the beginning of your experimental design.

In this example, ten families for each scenario have participated.

Table 6.2 illustrates an example of unbalanced data.

SAS's PROC ANOVA determines responses (# of trips per person) due to effects from the classification variables (homeownership and income level), and random error for the remaining variation. The example below illustrates how the procedure can be used.

Example: 6.8

Data: travel.xlsx

The travel data contains 3 variables: trips per person per day (trip), income (low, medium, and high), and homeownership status (yes and no).

The travel data is a set of balanced data.

trip	income	hos
5.2	low	yes
6.8	medium	yes
6.8	high	yes
4.9	low	no
6.4	medium	no
6.5	high	no
5.1	low	yes
6.6	medium	yes
6.6	high	yes
4.8	low	no
6.2	medium	no
6.3	high	no
5.2	low	yes
6.4	medium	yes
6.4	high	yes
4.6	low	no
6.1	medium	no
6.1	high	no
4.8	low	yes
6.2	medium	yes
6.2	high	yes
4.5	low	no
5.8	medium	no
5.9	high	no
4.7	low	yes
6.2	medium	yes
6.2	high	yes
4.3	low	no
5.7	medium	no
5.8	high	no
4.6	low	yes
5.8	medium	yes
5.8	high	yes
5.2	low	no
6.9	medium	no
6.2	high	no
5.5	low	yes
7.1	medium	yes
7.1	high	yes
6.4	low	no
7.3	medium	no
7.3	high	no
5.4	low	yes
7.1	medium	yes
7.1	high	yes
5.1	low	no
6.7	medium	no
6.8	high	no
5.3	low	yes
6.9	medium	yes
6.9	high	yes
6.2	low	no
8.1	medium	no
8.2	high	no
6.4	low	yes
7.3	medium	yes
7.3	high	yes
6.1	low	no
7.9	medium	no
7.1	high	no

Task: Determine how income and homeownership affect the daily trip rate

SAS Codes (E68)

```
TITLE "PROC ANOVA - Travel Data";
LIBNAME myt 'c:\sas_exercise';

PROC IMPORT OUT= WORK.travel
            DATAFILE= "c:\sas_
exercise\trip.xlsx"
            DBMS=EXCEL REPLACE;
     RANGE="trip$";
     GETNAMES=YES;
     MIXED=NO;
     SCANTEXT=YES;
     USEDATE=YES;
     SCANTIME=YES;
PROC ANOVA DATA=travel;
CLASS income hos;
MODEL trip = income hos income*hos;
RUN;
```

Explanations for codes from the TITLE statement until the PROC ANOVA; statement can be obtained from Section 3.5.1 Importing an Excel File.

PROC ANOVA invokes the ANOVA procedure for a data-set named travel.sas7bdat in the WORK library.

CLASS declares that income and hos are categorical variables.

*MODEL specifies that the trip is the dependent variable, and it may be affected by variables hos, income, and the interaction effects from hos*income.*

RUN instructs SAS to execute the PROC ANOVA procedure.

Result (E68):

1 ***PROC ANOVA - Travel Data***

2 ***The ANOVA Procedure***

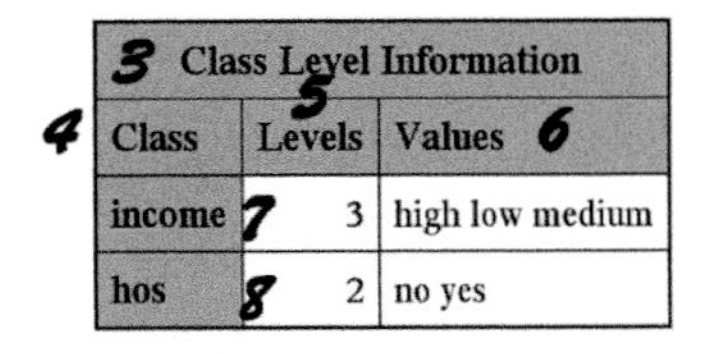

3 Class Level Information		
4 Class	5 Levels	Values 6
income	7 3	high low medium
hos	8 2	no yes

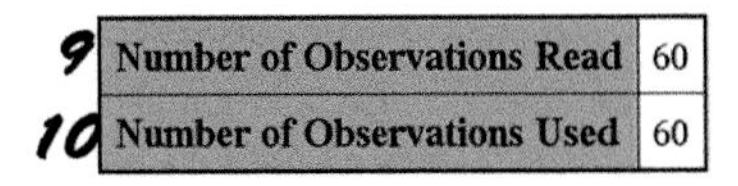

9 Number of Observations Read	60
10 Number of Observations Used	60

1 – it is the title specified.

2 – it is the default PROC ANOVA procedure subtitle.

3 – it offers information on the Class variables.

4 – under this column, it lists the income and hos variables.

5 – it shows that income has three levels and hos has two levels.

6 – it offers information on what the levels are.

7 and 8 – independent variables specified in the model.

9 and 10 – 60 are the number of observations (rows) read and used in the analysis, respectively. The Number of Observations Used may be fewer than the Number of Observations Read if missing values are encountered.

11 – it is the title specified.

12 – it is the default PROC ANOVA procedure subtitle.

13 – it identifies that the dependent variable is trip.

14 – this column outlines the sources of the model variation.

15 – degree of freedom.

11 PROC ANOVA - Travel Data

12 The ANOVA Procedure

Dependent Variable: trip trip 13

14 Source	15 DF	16 Sum of Squares	17 Mean Square	18 F Value	19 Pr > F
Model 20	5	27.59933333	5.51986667	12.95	<.0001
Error 21	54	23.01800000	0.42625926		
22 Corrected Total	59	50.61733333			

23 R-Square	24 Coeff Var	25 Root MSE	26 trip Mean
0.545255	10.57590	0.652885	6.173333

27 Source	28 DF	29 Anova SS	30 Mean Square	31 F Value	32 Pr > F
income	2	27.57233333	13.78616667	32.34	<.0001
hos	1	0.00266667	0.00266667	0.01	0.9372
income*hos	2	0.02433333	0.01216667	0.03	0.9719

16 – it refers to the Sum of Squares and has three parts. The first row is the Model SS. The second row is Error SS. And the last row is the Corrected total SS. Model SS = SUM $(Y_{model} - \bar{Y})^2$. Error SS = SUM$(Y - Y_{model})^2$. And the Corrected Total SS = SUM $(Y - \bar{Y})^2$.

17 – it is the result of dividing the Sum of Squares by the DF. It has two parts: Mean Square for the Model and Mean Square for the Error (Residual).

18 – it is the F statistics and computed by dividing the Model SS by the Error SS.

19 – it refers to the p value as related to the question of how reliable the independent variable predicts the dependent variable. In this example, the p value is < 0.0001. The hypothesis is rejected, and it is concluded that the independent variables do predict the dependent variable.

20 – this row shows measures as related to the model specified.

21 – this row shows measures as related to the error term.

22 – this row shows the combined amount from both the Model and Error components.

23 – it is the percentage of dependent variable variance that is predicted from the independent variables. In this example, it is 54.5255%. It reflects the explanation power of the entire model.

24 – it refers to the coefficient of variation measuring data variation. It is based on the formula of Coeff Var = (root mean square)/(mean of the dependent variable) × 100. The Coeff Var is unitless. In this example, it is 0.652885/6.1733333 × 100 = 10.57590.

25 – it refers to the standard deviation of the error term. It equals to the square root of the Mean Square for Error (Residual).

26 – 6.173333 is the mean of the dependent variable trip.

27 – this column lists all the independent variables and their interactions.

28 – it lists the Degree of Freedom for each Source.

29 – it is the total variance of an observed dataset. And it is calculated as SUM $(Y_i - \bar{Y})^2$.

30 – it is calculated based on the formula of SS/DF.

31 – it is the F statistic gained by dividing the Source Mean Square by the Error Mean Square for each Source (independent variable).

32 – it lists the p value for the null hypothesis (coefficient = 0) testing.

In this example, we should start with the two-way interaction effect between income and hos. The p value is 0.9719, which tells us that the null hypothesis can't be rejected. The interaction effect is not significant. Now, we can focus on the main effect. The p values for hos is 0.9372 and we failed to reject the hypothesis again and conclude that household size (hos) has no effect on the trip rate. The p value for income is <0.0001. The hypothesis is rejected, and we conclude that income has a significant effect on trip rate.

7

Visualization

In this chapter, for each example presented, I start with the data first. After the data is shown, I then present the relevant plots/charts/figures. The SAS code used to generate the plot/chart/figure is given last.

Using this sequence, you should have an easier time to learn the concept of SAS graphing. You should always first know your data, then formulate the type of graphic you are seeking. Only then should you start to actually put the SAS code together.

SAS's graphing capability is powerful. While basic graphing procedures are straightforward, they can get complicated easily and quickly. You can use my code as a starting point for your own specific tasks.

7.1 OVERVIEW

All examples presented in this chapter are done with the common graphing procedures of PROC GPLOT and PROC GCHART. As with virtually all SAS's PROC procedures, they can get the work done. However, the "option" specification can help you make a more tailored appearance. Options can be either global or local. A global option applies to all procedures in the SAS code file, whereas a local option is only applicable to a specific procedure. Local options override the prevailing global options.

To produce a chart or figure through SAS, you manipulate several key graphical components as listed below:

- Title/footnote/note
- Axis
- Legend
- Pattern
- Goptions.

The key is to understand how each of the components is specified (e.g., font, color, line thickness, order, sequence). Below is a summary overview of how to specify such basic graphical components.

7.1.1 Title, Footnote, and Note

The specification of title, footnote, and note for SAS's graphical procedure is the same as their specifications for all other SAS procedures.

While title and footnote statements are valid anywhere in a SAS code file, a note statement is only valid for the procedure it is specified under.

Below are example statements with different options.

TITLE Examples

```
TITLE1 FONT=swissb COLOR=black HEIGHT=14PT JUSTIFY=CENTER 'Age of
the Truck Fleet';
TITLE2 ANGLE=45 'Number of Vehicles';
TITLE3 COLOR=White BCOLOR=Black BOX=1 J =LEFT LSPACE=10 WRAP "Age
of the Truck Fleet";
TITLE4 LS=1.5 MOVE=(+10, +0) H=18PT ' Age of the Truck Fleet';
TITLE5 LS=1.0 MOVE=(+10,+0) H=16PT "Survey Data";
```

FOOTNOTE Examples

```
FOOTNOTE FONT=swiss COLOR=black HEIGHT=9PT JUSTIFY=CENTER
"Analyzed on 11/20/19";
FOOTNOTE2 ANGLE=45 '2019 Data';
FOOTNOTE3 COLOR=White BCOLOR=Black BOX=1 J =LEFT LSPACE=10 WRAP
"Purchased from CA";
FOOTNOTE4 LS=1.5 MOVE=(+10, +0) H=9PT 'AI Lab';
FOOTNOTE5 LS=1.0 MOVE=(+10,+0) H=9PT "College Park, MD";
```

NOTE Examples

NOTE statements are local options and need to be affiliated with a particular SAS procedure.

```
NOTE FONT=swiss COLOR=black HEIGHT=8PT JUSTIFY=CENTER "The 5th
Run";
NOTE Color=black MOVE=(60,60) BOX=2 "Survey Data of 2019";
```

7.1.2 Axis

Axes are the most critical part of a chart or figure. As a data analyst, you should have a good idea of what variables you are graphing. SAS GPLOT and GCHART procedures don't use the term *X*-axis or *Y*-axis. Instead, you encounter the following axis terminologies:

Haxis – horizontal axis
Vaxis – vertical axis
Raxis – response axis (typically refers to the vertical axis)
Maxis – midpoint axis (dealing with categorical variables, discrete values, and ranges)
Gaxis – group axis (dealing with groups with the BY statement)
Caxis – color of axis line, axis frame outline, and major/minor tick marks.

You customize the above axis (except Caxis) by linking each of them to a specific axis named as axisl to axisN first. For example, you can use the statements such as Haxis =axisl, Raxis=axis2, Maxis=axisl0, and Gaxis=14 to link the different axis to a numbered axis which you can specify further as illustrated below.

Case 1: Numerical and continuous axis data (e.g., speed in mph from 0 to 80)

SAS Codes (E712cl)

```
AXIS2
Value= (font=centbi
height=12pt color=black)
ORDER=(0 TO 80 BY 10)
MAJOR= (HEIGHT=12.5PT)
MINOR= (N=4 HEIGHT=6PT)
LABEL=(font=Swissb
HEIGHT=13pt ANGLE=90
JUSTIFY=C);
```

AXIS2 defines an axis named AXIS2 with characteristics as specified next.

The Value statement specifies that the black 12 pt centbi font is used with all values tied to the AXIS2.

The ORDER command specifies that the AXIS2 starts at 0 and ends at 80 with an increment of 10.

The MAJOR command specifies the major tick mark to be 12.5 pt in size.

The MINOR command instructs SAS to add 4 minor tick marks between each pair of major tick marks. And the MINOR tick mark is 6 pt in size.

The LABEL command specifies a title (either the variable name by default or something listed in quotation mark as in the E712c2 example) is used for AXIS2 and the 13 pt Swissb font should be used for the title. In addition, the title needs to be re-orientated (rotating) 90 degrees before centrally placed along the AXIS2.

Case 2: Categorical data for an axis (e.g., day of week)

SAS Codes (E712c2)

```
AXIS15
Value= (font=centbi
height=14pt color=black)
Order=('Monday'
'Tuesday' 'Wednesday'
'Thursday' 'Friday' 'Saturday' 'Sunday')
LABEL=(A=0 Height=11pt FONT=Swiss JUSTIFY=CENTER 'Day of Week');
```

AXIS15 defines an axis named AXIS15 with characteristics as specified next.

The Value statement specifies that the black 14 pt centbi font is used with all values associated with the AXIS15.

The Order command specifies that the AXIS15 is a discrete axis. It starts with Monday and ends with Sunday. Pay attention to the quotation marks used for each value.

The LABEL command specifies a title of Day of Week for the AXIS15 displayed with the black 11 pt Swiss font centrally.

Case 3: Categorical data for an axis (e.g., day of week) without additional label

SAS Codes (E712c3)

```
AXIS23
ORDER=('Sunday'
'Monday' 'Tuesday'
'Wednesday' 'Thursday'
'Friday' 'Saturday')
VALUE=(font= script
height=13pt
color=black)
LABEL=none
```

AXIS23 defines an axis named AXIS23 with characteristics as specified next.

The ORDER command specifies that the AXIS23 is a discrete axis and it starts with Monday and ends with Sunday. Pay attention to the quotation marks used for each value.

The VALUE statement specifies that the black 13 pt script font is used with all values associated with AXIS23.

The LABEL command specifies that AXIS23 has no title.

Case 4: Date axis (e.g., from a date to another by an interval of week)

SAS Codes (E712c4)

```
AXIS14
MAJOR=(HEIGHT=14 pt)
Order=("01SEP2019"d
TO "31DEC2019" d BY
WEEK)
LABEL=none;
```

AXIS14 defines an axis named AXIS14 with characteristics as specified next.

The MAJOR command specifies the major tick marks to be 14 pt in size.

The Order command specifies that the AXIS14 is a (d) date axis. It starts on September 1, 2019, and ends on December 31, 2019, with the step of WEEK.

The LABEL command specifies that AXIS14 has no title.

7.1.3 Legend

Legend differentiates data from different groups. It is always used with GROUP statements. The number of legends that a procedure generates and uses depends on the number of subgroups in a Group. For example, if your group of "highway" has three highways: I95, I5, and I80, your "Group by highway" statement will generate three legends: one for I95, one for I5, and one for I80.

A LEGEND statement (e.g., LEGEND=LEGENDn) controls the position and appearance of legends generated by the GCHART and GPLOT procedures. Example legend statements are illustrated below.

```
LEGEND1
LABEL="Highway"
COLOR=black
FONT=swissb
HEIGHT=12pt
JUSTIFY=Right
POSITION=(TOP CENTER INSIDE)
OFFSET=(0.1IN,0.15IN)
ORDER="I5" "I10" "I75"
;
```

7.1.4 Pattern

A pattern refers to how an area in a chart or plot is filled out. Such fill-outs include having solid fill, empty space, different colors, or different lines (slanting, crossing, and hatching).

A PATTERN statement provides controls on how an area is filled. Pattern statements are issued as PATTERN1, PATTERN2, PATTERN3, etc.

You specify a PATTERN statement by defining the following four parameters:

- Fill type: Value =
- Color type: Color=
- Image type: Image=
  (Image is used to fill a pattern in a 2D chart. For example, you can use a jpg file as the image to fill out an area).
- # of Repeats: Repeat =
  (number of times a pattern statement is repeated before using a new one).

Fill types (Value) are different for different graph types.

a. Value choices for bar and block charts under the PROC GCHART are as follows:
   - empty
   - solid
   - slanted lines (R, L, X) with different shading (1 to 5), where R stands for right slanted lines, L for left slanted lines, X for cross-hatched lines, and 1-5 characterize shading from the lightest 1 to the deepest of 5.

Examples

```
PATTERN1 VALUE=Empty COLOR=Gray;
PATTERN2 VALUE=SOLID COLOR=BLACK;
PATTERN3 V=L1 COLOR=BLACK;
PATTERN4 V=R5 COLOR=BLACK;
PATTERN5 V=R1 COLOR=BLACK;
PATTERN6 V=X4 COLOR=BLACK;
```

b. Value choices for a PIE chart under the PROC GCHART procedure are specified as:
   - PSOLID
   - PEMPTY
   - Shading from 1 to 5 with parallel lines (N), cross-hatched lines (X), and angle of pattern lines (0–360 degrees) specified together.

Examples

```
PATTERN1 VALUE=PSOLID COLOR=Gray;
PATTERN2 V=P1N45 COLOR=BLACK;
PATTERN3 V=P3N90 COLOR=BLACK;
PATTERN4 V=P53X135 COLOR=BLACK;
PATTERN5 V=P4X90 COLOR=BLACK;
```

7.1.5 GOPTIONS

The GOPTIONS refers to global graphical options and can be placed anywhere in your SAS code file. However, for any GOPTIONS to be effective, the options statements must be listed prior to your specific graphical procedures (e.g., PROC GPLOT) and executed prior to the specific graphical procedures.

Global options are extensive. Below is a list of examples related to text appearance (font, size, color).

It is a good practice to reset GOPTIONS for each of your procedure by using the statement below.

```
GOPTIONS RESET=GOPTIONS;
```

Below are common global options for graph size and text.

```
GOPTIONS
VSIZE=4IN
HSIZE=5.5IN

CBACK=white
CTEXT=black
CTITLE=Grey

FTEXT=swissb
FTITLE=arial

HTEXT=24PT
HTITLE=26PT;
```

7.2 PROC GPLOT

PROC GPLOT generates a plot of two or more variables on a single set of coordinate axes. The GPLOT procedure includes scatter plot, line plot, and bubble plot.

One of the best usages of the PROC GPLOT procedure is for generating a scatter plot. A scatter plot shows all data points illustrating the true nature of individual observations. PROC GPLOT procedure is unable to do computations (e.g., summarize, average). PROC GPLOT uses data as it is. You should prepare your data accordingly before using the procedure (e.g., compute the average, count, or summarize your data).

To illustrate this non-data computation limitation, let's say your data have repeated observations for a variable, and you want to plot the average of all the repeated values. The PROC GPLOT can't average these repeated observations for you. You need to get the average computed separately before the procedure can be used.

Example: 7.2

Data: speedplot.sas7bdat

The speedplot.sas7bdat data has eight variables: Road, SegID, SegLength, DOW, Time, WKWD, Speed, and Vehicles.

Vehicle speeds (**Speed**) at two different segments (**SegID**) of I-95 (**Road**) are measured by day of week (**DOW**) throughout a two-week period during both AM and PM (**Time**) peak commuting periods. The number of vehicles (**Vehicles**) possessing the measured speed during the time period is also recorded. A total of 28 observations (28 rows) are recorded. The full data are listed below.

Road	SegID	SegLength	DOW	Time	WKWD	Speed	Vehicles
I95	PG101	2.4	Monday	AM	weekday	35	3218
I95	PG101	2.4	Tuesday	AM	weekday	42	2690
I95	PG101	2.4	Wednesday	AM	weekday	32	3111
I95	PG101	2.4	Thursday	AM	weekday	58	2631
I95	PG101	2.4	Friday	AM	weekday	52	2860
I95	PG101	2.4	Saturday	AM	weekend	74	1989
I95	PG101	2.4	Sunday	AM	weekend	60	2421
I95	PG101	2.4	Monday	PM	weekday	36	3098
I95	PG101	2.4	Tuesday	PM	weekday	40	2987
I95	PG101	2.4	Wednesday	PM	weekday	28	3981
I95	PG101	2.4	Thursday	PM	weekday	36	3216
I95	PG101	2.4	Friday	PM	weekday	70	2122
I95	PG101	2.4	Saturday	PM	weekend	71	1898
I95	PG101	2.4	Sunday	PM	weekend	56	2122
I95	PG515	0.62	Monday	AM	weekday	39	3099
I95	PG515	0.62	Tuesday	AM	weekday	38	2591
I95	PG515	0.62	Wednesday	AM	weekday	36	2996
I95	PG515	0.62	Thursday	AM	weekday	24	3900
I95	PG515	0.62	Friday	AM	weekday	66	2754
I95	PG515	0.62	Saturday	AM	weekend	68	1916
I95	PG515	0.62	Sunday	AM	weekend	45	2332
I95	PG515	0.62	Monday	PM	weekday	41	2984
I95	PG515	0.62	Tuesday	PM	weekday	45	2877
I95	PG515	0.62	Wednesday	PM	weekday	25	3834
I95	PG515	0.62	Thursday	PM	weekday	52	2400
I95	PG515	0.62	Friday	PM	weekday	54	2044
I95	PG515	0.62	Saturday	PM	weekend	64	1828
I95	PG515	0.62	Sunday	PM	weekend	62	2044

Task: Create plots for speed by DOW by using different plot options

Case 1: Produce a speed scatter plot by day of week

Case 1: Result

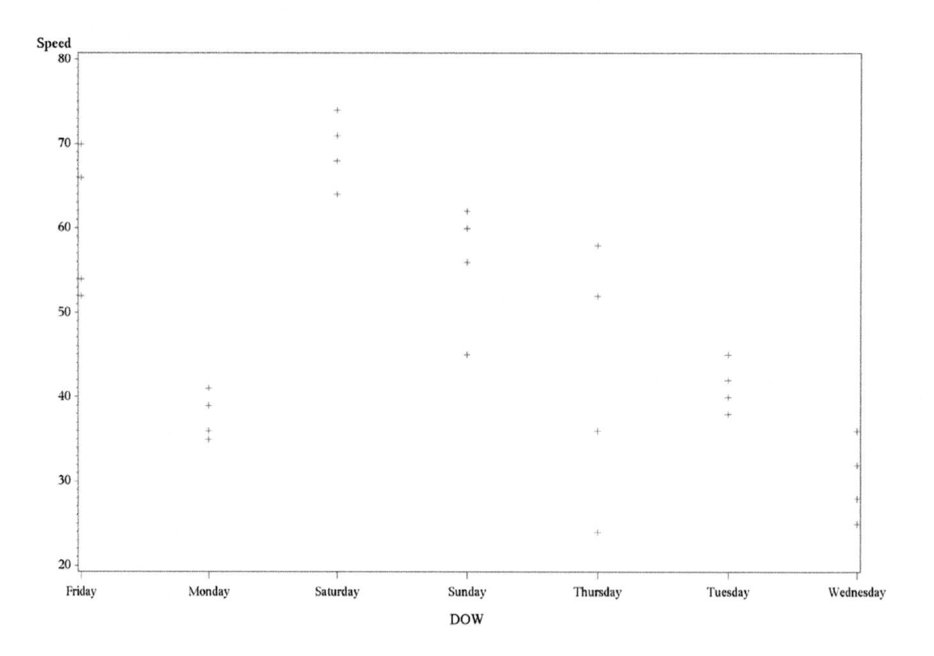

- The chart is a scatter plot where all the 28 speeds are plotted against the day of week it occurs.
- Each day of the week has four observations and is represented by the same blue+symbol.
- The chart's horizontal axis starts with Friday and ends with Wednesday, which is in alphabetical order based on the day of week spelling.
- The vertical axis is the speed which has the unit of mph. It starts at 20 mph and ends at 80 mph. It has a 10-mph major unit and 1-mph minor unit.
- No title.
- The label "Speed" for the vertical axis is at the top (not an ideal location).

Case 1: SAS Codes (E72c1)

* The plot is a simple default scatter chart. No options are used;

```
LIBNAME mydata 'c:\sas_exercise';
PROC GPLOT DATA=mydata.
myspeedplot;
PLOT Speed * DOW;
RUN;
QUIT;
```

LIBNAME defines a library called mydata representing the folder of "c:\sas_exercise."

PROC GPLOT invokes the GPLOT procedure for the dataset myspeedplot.sas7bdat in the mydata library.

*PLOT specifies that the variable Speed is to be plotted against the variable DOW. Pay attention to the * symbol used between Speed and DOW.*

RUN instructs SAS to execute the PROC GPLOT procedure.

QUIT terminates the procedure.

Case 2: Produce a speed scatter plot by day of week with AM and PM differentiation

Case 2: Result

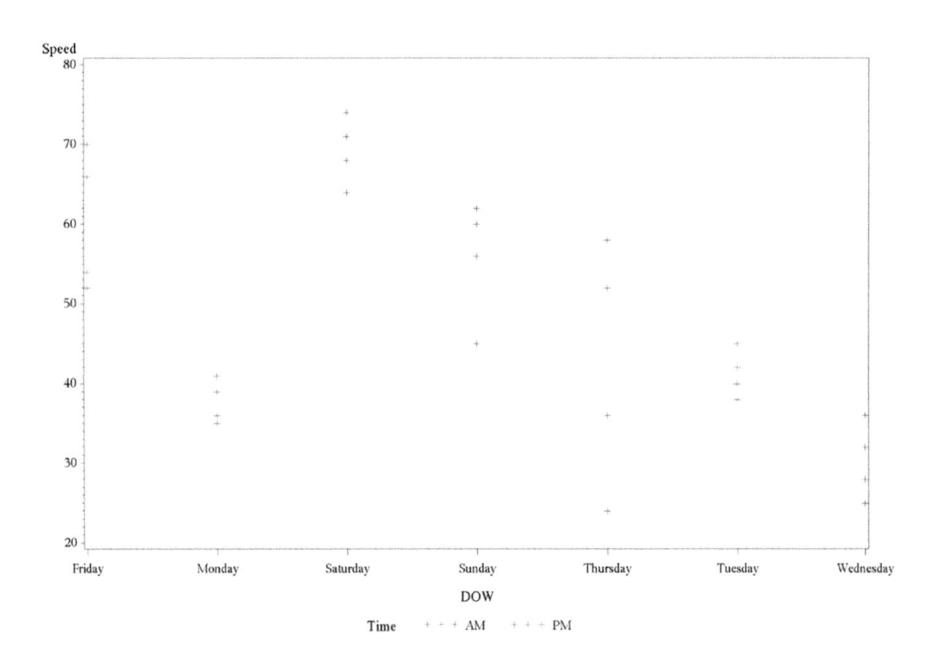

- Similar to the Case 1 example except that the AM and PM observations are identified with different symbols: AM is represented by a blue+sign, and PM is represented by a red+sign.

Case 2: SAS Codes (E72c2)

*Time (AM and PM) is differentiated by default color difference for symbols;

```
LIBNAME mydata
'c:\sas_exercise';
PROC GPLOT DATA=mydata.
myspeedplot;
PLOT Speed * DOW = time;
RUN;
QUIT;
```

LIBNAME defines a library called mydata representing the folder of "c:\sas_exercise."

PROC GPLOT invokes the GPLOT procedure for the dataset myspeedplot.sas7bdat in the mydata library.

*PLOT specifies that the variable Speed is to be plotted against the variable DOW and differentiated by the variable time. Pay attention to the * and=symbols used.*

RUN instructs SAS to execute the PROC GPLOT procedure.

QUIT terminates the procedure.

Case 3: Produce a speed scatter plot by day of week, but with AM from PM observations differentiated with more obviously different symbols.

Case 3: Result

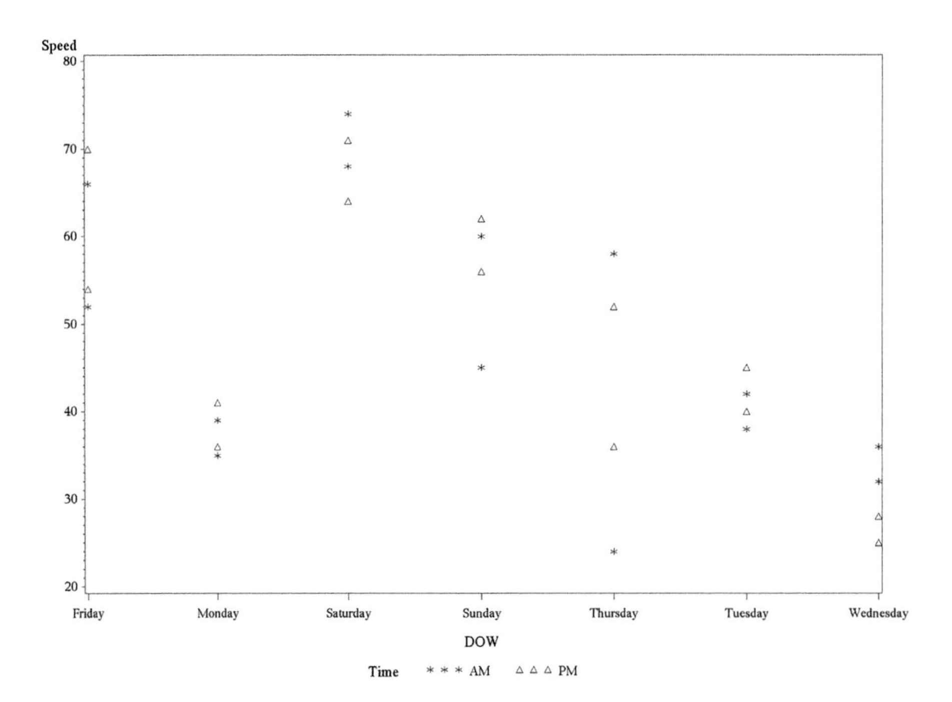

- Similar to the Case 2 example except that the AM and PM observations are identified with new symbols: AM is represented by black *, and PM is represented by black triangle Δ.

Case 3: SAS Codes (E72c3)

*Time (AM and PM) is differentiated by customized symbol statements;

```
GOPTIONS RESET=GLOBAL;
SYMBOL1 VALUE=STAR
COLOR=BLACK ;
SYMBOL2 VALUE=TRIANGLE
COLOR=BLACK ;
LIBNAME mydata
'c:\sas_exercise';
PROC GPLOT DATA=mydata.
myspeedplot;
PLOT Speed*DOW=time;
RUN;
QUIT;
```

GOPTIONS is the command to declare graphical options. RESET=GLOBAL instructs SAS to reset all graphical options to default.

SYMBOL1 is an SAS command that specifies the first symbol (if needed by any SAS procedure) is a star (VALUE=STAR) and has the color of black (COLOR=BLACK).

SYMBOL2 is an SAS command that specifies the second symbol (if needed by any SAS procedure) is a triangle (VALUE=TRIANGLE) and has the color of black (COLOR=BLACK).

LIBNAME defines a library called mydata representing the folder of "c:\sas_exercise."

PROC GPLOT invokes the GPLOT procedure for the dataset myspeedplot.sas7bdat in the mydata library.

*PLOT specifies that the variable Speed is to be plotted against the variable DOW and differentiated by the variable time (SYMBOL1 and SYMBOL2 are used to differentiate time). Pay attention to the * and=symbols used.*

RUN instructs SAS to execute the PROC GPLOT procedure.

QUIT terminates the procedure.

TABLE 7.1

Some Common SAS Plot Symbol Values

Value=	Symbol
A	A
Plus	+
X	X
Star	*
Dot	•
Diamond	◊
Square	□
Triangle	△
=	☆
#	♡
Circle	○

Case 4: Produce a journal quality speed scatter plot by day of week

Case 4: Result

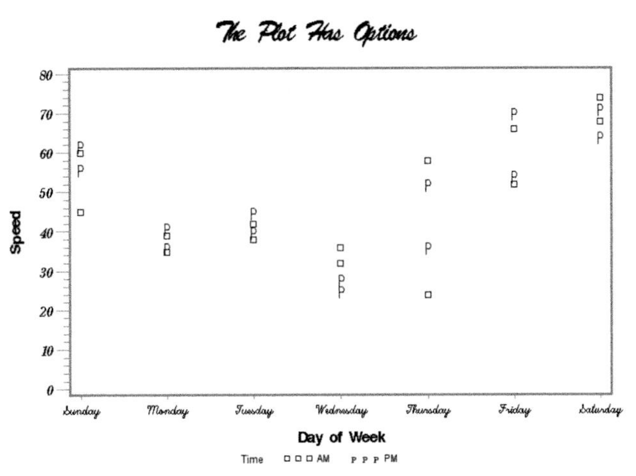

- The above chart is a highly customized scatter plot.
- It has a title and a footnote.
- The horizontal axis (defined as axis1) has the day of week arranged in a sequential order started with Sunday with the "Day of Week" label.
- The vertical axis (defined as axis2) has a new scale from 0 to 80. The main tick mark is 10 mph apart. The minor tick mark has a unit of 2 mph (four ticks within a pair of main tick marks).
- Symbols for the AM and PM are different and easy to distinguish.
- The label "Speed" for the vertical axis (axis2) is center-adjusted with its own font and size.

Case 4: SAS Codes (E72c4)

```
GOPTIONS RESET=GLOBAL;

TITLE HEIGHT=24PT COLOR=BLACK FONT=Brush 'The Plot Has Options'   ;
FOOTNOTE1 HEIGHT=11PT JUSTIFY=LEFT FONT= simplex 'Source: CK';

SYMBOL1 VALUE=SQUARE COLOR=BLACK HEIGHT=10PT ;
SYMBOL2 VALUE='P' COLOR=BLACK HEIGHT=14PT;

AXIS1 ORDER=('Sunday' 'Monday' 'Tuesday' 'Wednesday' 'Thursday'
'Friday' 'Saturday')
      VALUE=(font= script height=13pt color=black)
      LABEL=(font= swissb HEIGHT=13pt COLOR=black 'Day of Week')
      OFFSET=(10PT,10PT)
      WIDTH=2;
AXIS2
Value= (font=centbi height=12pt color=black)
ORDER=(0 TO 80 BY 10)
MAJOR= (HEIGHT=12.5PT)
MINOR= (N=4 HEIGHT=6PT)
LABEL=(font=Swissb HEIGHT=13pt ANGLE=90 JUSTIFY=C);

LIBNAME mydata 'c:\sas_exercise';
PROC GPLOT DATA=mydata.myspeedplot;
PLOT Speed*DOW=time/HAXIS=axis1 VAXIS=axis2;
RUN;
QUIT;
```

GOPTIONS is the command to declare graphical options. RESET=GLOBAL instructs SAS to reset all graphical options to default.

The TITLE lists all title-related specifications:

HEIGHT= 24PT (title font size)
COLOR=BLACK (font color)
FONT=Brush (font type)
"The Plot Has Options" (actual title).

The FOOTNOTE1 lists footnote-related specifications:

HEIGHT=11PT (font size)
JUSTIFY=LEFT (footnote location justification, left, right, or center)
FONT=simplex (font type)
"Source: CK" (actual footnote).

SYMBOL1 lists symbol1-related specifications:

VALUE=SQUARE (The type of symbol to use)
COLOR=BLACK (symbol color)
HEIGHT=10PT (font size).

SYMBOL2 lists symbol2-related specifications:

VALUE= "p" (The letter p is used as a symbol)
COLOR=BLACK (symbol color)
HEIGHT=14PT (font size).

AXIS1 lists its specifications as follows:

ORDER=('Sunday' … ' 'Saturday'). Items for AXIS1 from left to right in a chronological order.
VALUE=…. font, height, and color
LABEL =(font= swissb HEIGHT=13pt COLOR=black "Day of Week"). The title for AXIS1 is Day of Week with the black 13 pt swissb font
OFFSET=(10PT,10PT) tells SAS to shift the AXIS1 starting point 10 points up and 10 points right from the default location
WIDTH=2 is a factor defining the width of the plot frame.

AXIS2 lists its specifications as follows:

Value= (font=centbi height=12pt color=black) specifies font, size, and color.
ORDER= (0 TO 80 BY 10) instructs SAS to start at 0 and end at 80 with an increment of 10.
MAJOR=(HEIGHT=12.5PT) instructs SAS to use 12.5 pt major tick marks.
MINOR=(N=4 HEIGHT=6PT) instructs SAS to insert 4 minor tick marks between each pair of major tick marks with a height of 6 points.
LABEL=(font=Swissb HEIGHT=13pt ANGLE=90 JUSTIFY=C) defines AXIS2 title to be displayed with 13 pt Swissb font and centrally displayed after rotating clockwise 90 degrees from its normal position.

LIBNAME defines a library called mydata representing the folder of "c:\sas_exercise."

PROC GPLOT invokes the GPLOT procedure for the dataset myspeedplot.sas7bdat in the mydata library.

*PLOT specifies that variable Speed is to be plotted against the variable DOW and differentiated by the variable time. Pay attention to the usage of the * symbol.*

The /HAXIS=axis1 VAXIS=axis2 are optional commands declaring that the horizontal axis (HAXIS) is axis1 and the vertical axis (VAXIS) is axis2.

RUN instructs SAS to execute the PROC GPLOT procedure.

QUIT terminates the procedure.

Case 5: Produce journal quality scatter plot comparing speed and number of vehicles (quantitative data for both axes).

The example plot shows the speed observed (vertical axis) and the corresponding number of vehicles observed (horizontal axis) for that speed.

Case 5: Result

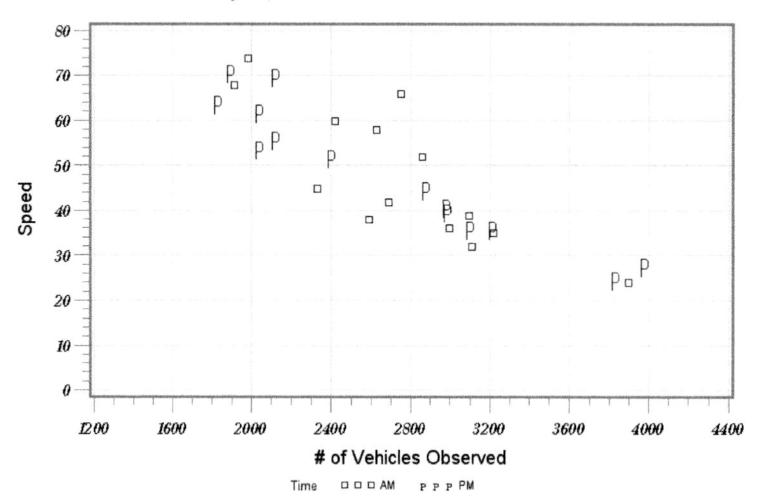

Source: CT

- The chart plots the speed on the vertical axis and the number of corresponding vehicles on the horizontal axis.
- It has both a title and a footnote. Notice location differences for them as compared to the last example (Case 4).
- The horizontal axis (defined as axis1) now has a scale starting from 1200 and ending at 4400 with 400 increment per major tick marks. Within a pair of major tick marks, there are three minor tick marks dividing them and making the minor unit to be 100.
- The vertical axis (defined as axis2) has a new scale from 0 to 80 with 10 mph as a major unit. Each 10 mph is divided into five subunits of 2 mph each (four minor ticks within a pair of main tick marks).
- Symbols for the AM and PM are very different and easy to distinguish.
- The label for the vertical axis (axis2) is center-adjusted with its own font and size.
- It has both vertical and horizontal reference lines.

Case 5: SAS Code (E72c5)

```
GOPTIONS RESET=GLOBAL;
TITLE HEIGHT=24PT COLOR=BLACK FONT=Brush JUSTIFY=LEFT 'The Plot
Has Lots of Options'  ;
```

```
FOOTNOTE1 HEIGHT=12PT JUSTIFY=Right FONT = simplex 'Source: CT';

SYMBOL1 VALUE=SQUARE COLOR=BLACK HEIGHT=10PT ;
SYMBOL2 VALUE='P' COLOR=BLACK HEIGHT=18 POINT;

AXIS1
      ORDER=(1200 to 4400 by 400)
      Value= (font=centbi height=12pt color=black)
      MAJOR= (HEIGHT=12.5PT)
      MINOR= (N=3 HEIGHT=6PT)
      WIDTH=2
      LABEL= (font=centbi HEIGHT=15pt COLOR=black '# of Vehicles
Observed');

AXIS2
      ORDER=(0 TO 80 BY 10)
      Value= (font=centbi height=12pt color=black)
      MAJOR= (HEIGHT=12.5PT)
      MINOR= (N=4 HEIGHT=6PT)
      LABEL=(HEIGHT=15 PT ANGLE=90 JUSTIFY=C);

LIBNAME mydata 'c:\sas_exercise';
PROC GPLOT DATA=mydata.myspeedplot;
PLOT Speed*vehicles=time/HAXIS=axis1 VAXIS=axis2
AUTOHREF
AUTOVREF;
RUN;
QUIT;
```

GOPTIONS is the command to declare graphical options. RESET=GLOBAL instructs SAS to reset all graphical options to default.

The TITLE lists all title-related specifications:

HEIGHT= 24PT (title font size)
COLOR=BLACK (font color)
FONT=Brush (font type)
JUSTIFY=LEFT (left, right, center justification)
"The Plot Has Lots of Options" (actual title).

The FOOTNOTE1 lists footnote-related specifications:

HEIGHT=12PT (font size)
JUSTIFY=RIGHT (footnote location justification, left, right, or center)
FONT=simplex
"Source: CT" (actual footnote)

SYMBOL1 lists symbol1-related specifications:

VALUE=SQUARE (types of symbols to use)
COLOR=BLACK (symbol color)

HEIGHT=10PT (font size).
SYMBOL2 lists symbol2-related specifications:
VALUE= "p" (types of symbols to use. Here, the letter P is used as a symbol)
COLOR=BLACK (symbol color)
HEIGHT=18 POINT (font size).
AXIS1 lists all its specifications as follows:
ORDER=(1200 to 4400 by 400) instructs SAS to start AXIS1 at 1200 and end at 4400 with an increment of 400.
Value= (font=centbi height=12pt color=black) specifies the font, color, and size.
MAJOR=(HEIGHT=12.5PT) instructs SAS to have the major tick mark with a height of 12.5 points.
MINOR=(N=3 HEIGHT=6PT) instructs SAS to insert 3 minor tick marks between each pair of major tick marks with a height of 6 points.
LABEL=(font=centbi HEIGHT=15pt COLOR=black) tells SAS to use black 15 pt centbi font for Axis1 title of "# of Vehicles Observed."
AXIS2 lists its specifications as follows:
ORDER =(0 TO 80 BY 10) instructs SAS to start AXIS2 at 0 and end at 80 with an increment of 10.
VALUE= (FONT=CENTBI HEIGHT=12PT COLOR=BLACK) tells SAS to use the black 12 pt Centbi font to display all AXIS2 values.
MAJOR= (HEIGHT=12.5PT) instructs SAS to have the major tick mark with a height of 12.5 points.
MINOR= (N=4 HEIGHT=6PT) instructs SAS to insert 4 minor tick marks between each pair of major tick marks with a height of 6 points.
LABEL=(HEIGHT=15PT ANGLE=90 JUSTIFY=C) defines AXIS2 label will be 15 pt high, rotated clockwise 90 degrees from its normal position, and center-justified. No specific label is listed by the LABEL command. Consequently, the default variable "Speed" is used as the title for AXIS2.
LIBNAME defines a library called mydata representing the folder of "c:\sas_exercise."
PROC GPLOT invokes the GPLOT procedure for the dataset myspeedplot.sas7bdat in the mydata library.
*PLOT specifies that variable Speed is to be plotted against the variable vehicles differentiated by the variable time. Pay attention to the usage of the * symbol*
The /HAXIS=axis1 VAXIS=axis2 are optional commands declaring that the horizontal axis is axis1 and the vertical axis is axis2.
AUTOHREF invokes the plotting of horizontal (H) reference (REF) lines at every major tick marks.
AUTOVREF invokes the plotting of vertical (V) reference (REF) lines at every major tick marks.
RUN instructs SAS to execute the PROC GPLOT procedure.
QUIT terminates the procedure.

Case 6: Produce a journal quality bubble chart where speed, DOW, and vehicles are all illustrated.

A bubble chart with a vertical axis for speed, horizontal axis for DOW, and varying bubble sizes proportional to the number of vehicles is desired.

Unlike a line chart and a scatter plot that only handle two-dimensional data, a bubble chart offers the ability to display a third-dimensional parameter (variable) through the size of the bubble. For example, you can present information on vehicle speed by day of the week and the number of vehicles observed in a single bubble chart.

Case 6: Result

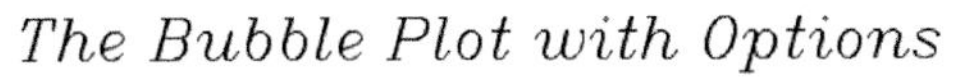

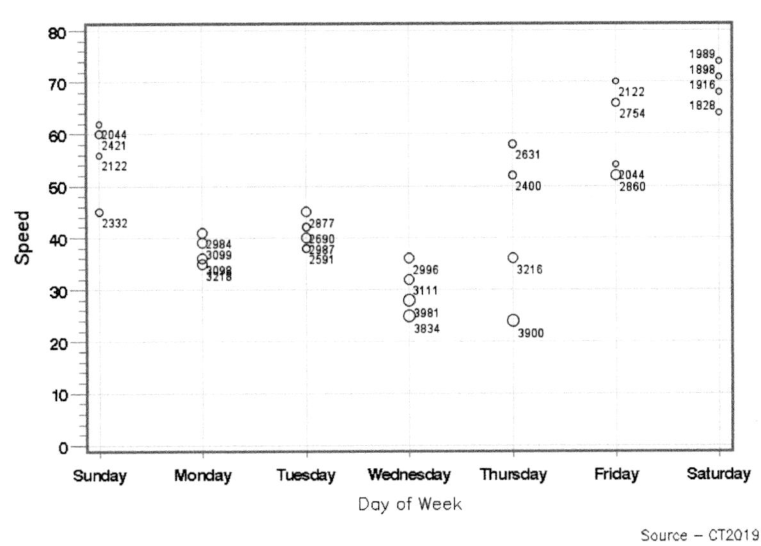

- The chart presents data in a three-dimensional manner: speed, DOW, and the number of vehicles.
- The horizontal axis is Day of Week, the vertical axis is speed, and the bubble (circle) size is proportional to the number of vehicles observed for that speed.
- The chart is journal quality with all needed attributes for journal publication.

Case 6: SAS Codes (E72c6)

***Journal quality bubble chart with a lot of options;**

```
GOPTIONS RESET=GLOBAL;

TITLE HEIGHT=24PT COLOR=BLACK FONT=Swissbi JUSTIFY=CENTER 'The
Bubble Plot with Options' ;
FOOTNOTE HEIGHT=12PT JUSTIFY=Right FONT = simplex 'Source
- CT2019';

AXIS1 ORDER=('Sunday' 'Monday' 'Tuesday' 'Wednesday' 'Thursday'
'Friday' 'Saturday')
      VALUE=(FONT=SWISSB COLOR=black HEIGHT=12PT )
```

```
      LABEL = (FONT= simplex COLOR=black  HEIGHT=15PT  'Day of
Week')
      WIDTH=2
      OFFSET=(11PT,12PT);

AXIS2 ORDER =(0 TO 80 BY 10)
        VALUE=(HEIGHT=12PT FONT=SWISSB COLOR=BLACK)
        MAJOR= (HEIGHT=12.5PT)
        MINOR= (N=4 HEIGHT=6PT)
      LABEL=(HEIGHT=15pt ANGLE=90 JUSTIFY=C);

LIBNAME mydata 'c:\sas_exercise';

PROC GPLOT DATA=mydata.myspeedplot;
BUBBLE Speed*DOW=vehicles/HAXIS=axis1 VAXIS=axis2
BSCALE=RADIUS
BSIZE=20
BCOLOR=black
BLABEL
AUTOHREF
AUTOVREF;
RUN;
QUIT;
```

GOPTIONS is the command to declare graphical options. RESET=GLOBAL instructs SAS to reset all graphical options to default.

The TITLE lists title-related specifications:

HEIGHT=24PT (title font size)
COLOR=BLACK (font color)
FONT=Swissbi (font type)
JUSTIFY=CENTER (left, right, or center justification)
"The Bubble Plot with Options" (actual title).

The FOOTNOTE lists all footnote-related specifications:

HEIGHT=12PT (font size)
JUSTIFY=Right (footnote location justification, left, right, or center)
FONT=simplex
"Source - CT2019" (actual footnote)

AXIS1 lists all its unique specifications as follows:

ORDER=('Sunday'.... 'Saturday') arranges the sequence of the AXIS1 values to be displayed from left to right. It starts with Sunday, then Monday … and ends with Saturday in a chronological order

VALUE=(FONT=SWISSB COLOR=black HEIGHT=12PT) tells SAS to use the black 12 pt SWISSB font to display all AXIS1 values.

LABEL=(FONT=simplex COLOR=black HEIGHT=15PT) instructs SAS to title the AXIS1 with "Day of Week" by using black 15 pt simplex font.

WIDTH=2 specifies that the plot frame width has a factor of 2.

OFFSET=(11PT,12PT) tells SAS to shift the AXIS1 starting point 11 points up and 12 points right from its default location.

AXIS2 lists all its unique specifications as follows:

ORDER=(0 TO 80 BY 10) instructs SAS to start AXIS2 at 0 and end at 80 with an increment of 10.

VALUE=(HEIGHT=12PT FONT=SWISSB COLOR=BLACK) tells SAS that the all numerical numbers associated with AXIS2 should use the black 12 pt, SWISSB font.

MAJOR=(HEIGHT=12.5PT) instructs SAS to have all major tick marks with a height of 12.5 points.

MINOR=(N=4 HEIGHT=6PT) instructs SAS to insert 4 minor tick marks between each pair of major tick marks with a height of 6 points.

LABEL=(HEIGHT=15pt ANGLE=90 JUSTIFY=C) defines AXIS2 title will be 15 pt high, rotated clockwise 90 degrees from its normal position, and center-justified.

LIBNAME defines a library called mydata representing the folder of "c:\sas_exercise."

PROC GPLOT invokes the GPLOT procedure for the dataset myspeedplot.sas7bdat in the mydata library.

*BUBBLE Speed*DOW=vehicles specifies that the variable Speed is to be plotted against the variable DOW and further differentiated by the variable vehicles through bubble sizes. Pay attention to the usages of the * and=symbols.*

The /HAXIS=axis1 VAXIS=axis2 are optional commands declaring that the horizontal axis is axis1 and the vertical axis is axis2.

BSCALE=RADIUS tells SAS how to differentiate bubble sizes - by RADIUS or surface AREA. In this example, a bubble RADIUS is used to represent the number of vehicles the bubble represents.

BSIZE =20 is a scaling factor that can be adjusted to make the bubble fit your overall layout appropriately.

BCOLOR =black instructs SAS to draw the bubble in black.

BLABEL asks SAS to label a bubble with the actual value of vehicles. If you do not want to label your bubble, just remove the BLABEL.

AUTOHREF invokes the plotting of horizontal (H) reference (REF) lines at every major tick marks.

AUTOVREF invokes the plotting of vertical (V) reference (REF) lines at every major tick marks.

RUN instructs SAS to execute the PROC GPLOT procedure.

QUIT terminates the procedure.

Case 7: Plot of Summary Data

The speedplot.sas7bdat data contains summary data items, including average speed (AMSpeed, PMSpeed, and ASpeed) and the average number of vehicles (AMvehicle, PMvehicle, and Avehicle) observed by time (AM, PM, and average of AM and PM) and day of the week (Sunday ... Saturday). The data is presented below.

Data: average_speed.sas7bdat

DOW	AMspeed	PMSpeed	ASpeed	AMvehicle	PMvehicle	Avehicle
Friday	59	62	60.5	2807	2083	2445
Monday	37	38.5	37.75	3158.5	3041	3099.75
Saturday	71	67.5	69.25	1952.5	1863	1907.75
Sunday	52.5	59	55.75	2376.5	2083	2229.75
Thursday	41	44	42.5	3265.5	2808	3036.75
Tuesday	40	42.5	41.25	2640.5	2932	2786.25
Wednesday	34	26.5	30.25	3053.5	3907.5	3480.5

Case 7A: Illustrate how to plot AMspeed and PMSpeed against DOW on a single chart by overlaying the two plots together (AMspeed and PMSpeed have the same scale of magnitude)

Case 7A Result: Overlay plot with similar scale of Y variables

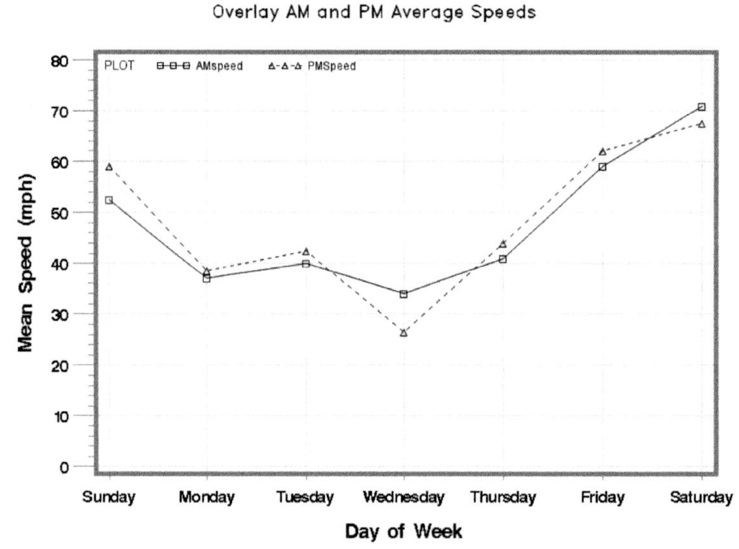

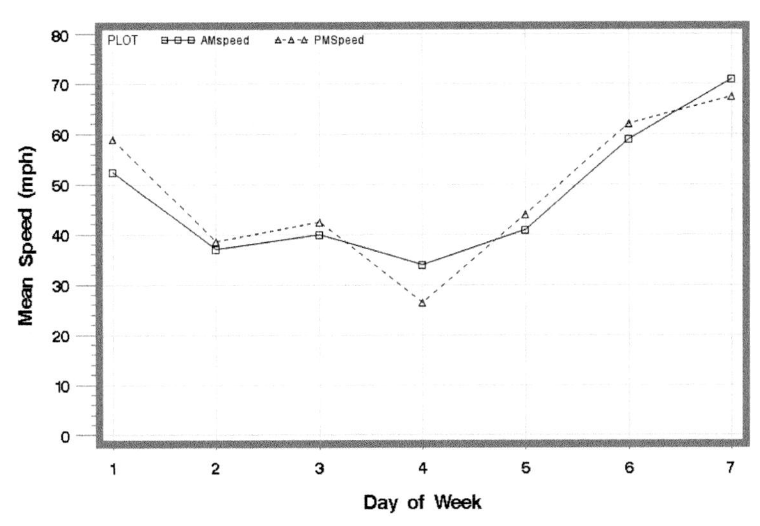

- The chart is journal quality and ready to be used for publication.
- The chart provides a clear pattern on how AM and PM average speeds change as the day moves systematically up in a week. The seven points are connected by a straight line.
- Both the AMspeed and PMSpeed use the same left-side vertical axis (same scale).

Case 7A: SAS Codes (E72c7a)

```
GOPTIONS RESET=GLOBAL;

TITLE HEIGHT=16PT COLOR=BLACK FONT=Duplex 'Overlay AM and PM
Average Speeds'  ;
FOOTNOTE1 HEIGHT=12PT JUSTIFY=LEFT 'Pay Attention to DOW and
DOWID';

SYMBOL1 VALUE=SQUARE COLOR=BLACK HEIGHT=10PT I=join L= 1;
SYMBOL2 VALUE=Triangle COLOR=BLACK HEIGHT=10PT I=join L=2;

AXIS1 ORDER=('Sunday' 'Monday' 'Tuesday' 'Wednesday' 'Thursday'
'Friday' 'Saturday' )
      VALUE=(FONT=swissb color=black HEIGHT=12PT )
      LABEL =(FONT=swissb HEIGHT=15pt  COLOR=black 'Day of Week')
      OFFSET=(10PT,10PT)
      WIDTH=4;

AXIS2 ORDER =(0 TO 80 BY 10)
        VALUE=(HEIGHT=12PT FONT=swissb)
        MAJOR= (HEIGHT=18PT)
        MINOR= (N=4 HEIGHT=6PT)
```

```
      LABEL=(FONT=swissb Color=black HEIGHT=15pt ANGLE=90
JUSTIFY=C 'Mean Speed (mph)');

AXIS3 ORDER =(1 TO 7 BY 1)
      VALUE=(FONT=swissb Color =black HEIGHT=12PT)
        LABEL =(FONT=swissb HEIGHT=15pt  COLOR=black 'Day of
Week')
      OFFSET=(10PT,10PT)
      WIDTH=7;

LEGEND1     POSITION=(top left inside);

LIBNAME mydata 'c:\sas_exercise';

DATA mydata.as2;
set mydata.average_speed;
IF DOW='Sunday' THEN DOWID=1 ;
IF DOW ='Monday' THEN DOWID=2 ;
IF DOW='Tuesday' THEN DOWID=3;
IF DOW ='Wednesday' THEN DOWID=4;
IF DOW='Thursday' THEN DOWID=5;
IF DOW ='Friday' THEN DOWID=6;
if DOW='Saturday' THEN DOWID=7;
RUN;

PROC Sort data=mydata.as2;
BY DOWID;
RUN;


PROC GPLOT DATA=mydata.as2;
PLOT AMSpeed*DOW PMspeed*DOW /OVERLAY HAXIS=axis1 VAXIS=axis2
LEGEND=LEGEND1
AUTOHREF
AUTOVREF;
RUN;
Quit;

PROC GPLOT DATA=mydata.as2;
PLOT AMSpeed*DOWID PMspeed*DOWID /OVERLAY HAXIS=axis3 VAXIS=axis2
LEGEND=LEGEND1
AUTOHREF
AUTOVREF ;
RUN;
QUIT;
```

GOPTIONS is the command to declare graphical options. RESET=GLOBAL instructs SAS to reset all graphical options to default.

The TITLE lists title-related specifications:

HEIGHT= 16PT (title font size)
COLOR=BLACK (font color)
FONT=Duplex (font type)
"Overlay ....." (actual title).

The FOOTNOTE1 lists all footnote-related specifications:

HEIGHT=12PT (font size)
JUSTIFY=LEFT (footnote location justification, left, right, or center)
"Pay Attention to DOW and DOWID" (actual footnote)

SYMBOL1 specifies that the first symbol (if needed by any SAS procedure) is a square (VALUE=SQUARE), having the color of black (COLOR=BLACK), size of 10 pt (HEIGHT=10PT), simple joint (I=join), and line style of 1 (L=1).

SYMBOL2 specifies that the second symbol (if needed by any SAS procedure) is a triangle (VALUE=Triangle), having the color of black (COLOR=BLACK), size of 10 pt (HEIGHT=10PT), simple joint (I=join), and line style of 2 (L=2).

AXIS1 lists its specifications as follows:

ORDER=('Sunday'.... 'Saturday') arranges the sequence of AXIS1 values to be displayed from left to right. It starts with Sunday, then Monday ... and ends with Saturday.
VALUE=(FONT=swissb color=black HEIGHT=12PT) tells SAS to use the black 12 pt swissb font for values displayed along AXIS1.
LABEL =(FONT=swissb HEIGHT=15pt COLOR=black) defines the AXIS1 title to be "Day of Week" and displayed with the black 15 pt swissb font.
OFFSET=(10PT,10PT) tells SAS to shift the AXIS1 starting point 10 points up and 10 points right from the default location.
WIDTH=4 specifies that the plot frame width has a factor of 4.

AXIS2 lists its specifications as follows:

ORDER=(0 TO 80 BY 10) instructs SAS that AXIS2 starts at 0 and ends at 80 with an increment of 10.
VALUE=(HEIGHT=12PT FONT=swissb) tells SAS to use the 12 pt swissb font for all the AXIS2 numerical numbers.
MAJOR=(HEIGHT=18PT) instructs SAS to have the major tick mark with a height of 18 points.
MINOR=(N=4 HEIGHT=6PT) instructs SAS to insert 4 minor tick marks between each pair of major tick marks with a height of 6 points.
LABEL=(FONT=swissb Color=black HEIGHT=15pt ANGLE=90 JUSTIFY=C 'Mean Speed (mph)') defines "Mean Speed (mph)" as the AXIS2 title and the black 15 pt swissb font is used for the title display. The title should be rotated 90 degrees clockwise before centrally placed.
AXIS3 ORDER =(1 TO 7 BY 1) defines AXIS3 to start at 1 and end at 7 with 1 as the step increment.
VALUE=(FONT=swissb Color =black HEIGHT=12PT) specifies that the black 12 pt swissb font is to be used.
LABEL =(FONT=swissb HEIGHT=15pt COLOR=black 'Day of Week') instructs SAS to use black 15 point swissb font to display the title Day of Week.

OFFSET=(10PT,10PT) tells SAS to shift the AXIS3 starting point 10 points up and 10 points right from the default location.
WIDTH=7 specifies that the plot frame width has a factor of 7.
LEGNED1 POSITION= (top left inside) directs SAS to put the legends of AMspeed and PMspeed at the top left and within the plot frame.
LIBNAME defines a library called mydata representing the folder of "c:\sas_exercise."
DATA mydata.as2 defines a library named mydata representing the folder of c:\ sas_exercise.
set mydata.average_speed tells SAS to read a data file named average_speed.sas7bdata in the mydata folder.
IF DOW="Sunday" THEN DOWID=1 creates a new variable called DOWID and assigns a value of 1 if the condition is met.
IF DOW … THEN … fills the observations of all DOWID, respectively, to the corresponding DOW value.
PROC Sort data=mydata.as2 invokes the sort procedure for data as2 in the mydata library. Sorting is done by DOWID.
PROC GPLOT invokes the GPLOT procedure for the dataset as2.sas7bdat in the mydata library.
*PLOT AMSpeed*DOW PMspeed*DOW /OVERLAY HAXIS=axis1 VAXIS=axis2 instructs SAS to plot the AMSpeed against DOW, plot PMspeed against DOW, and then overlay them on top of each other. It further specifies that the horizontal axis is axis1 and the vertical axis is axis2.*
LEGEND=LEGEND1 declares that LEGEND used should follow what is specified by LEGEND1.
AUTOHREF draws the plotting of horizontal (H) reference (REF) lines at every major tick marks on the response axis (RAXIS).
AUTOVREF draws the plotting of horizontal (H) reference (REF) lines at every major tick marks on the response axis (RAXIS).
RUN instructs SAS to execute the procedure.
Quit terminates the procedure.

Case 7B: Illustrate how to plot AMspeed and AMvehicle against DOW on a single chart through dual Y-axis (Vehicles and AMspeed have totally different magnitude)

Case 7B: Result

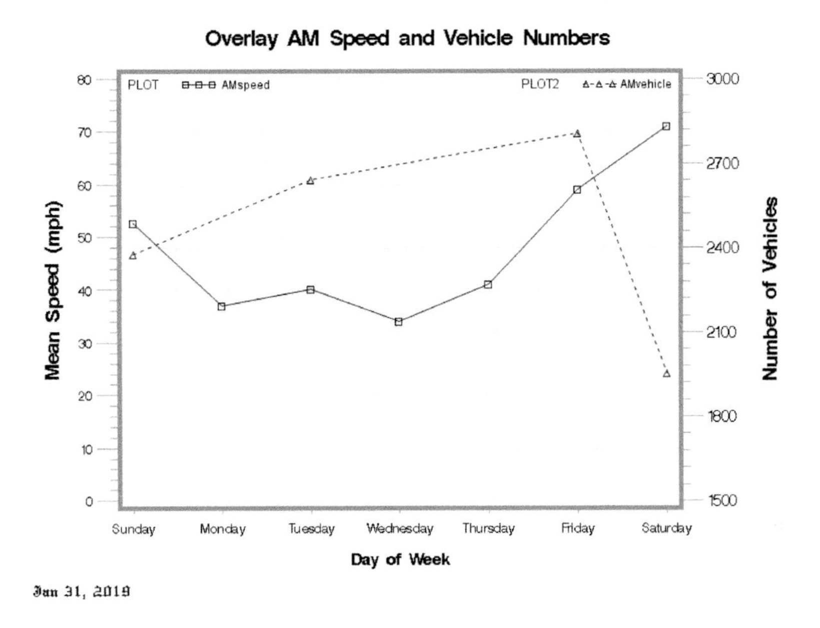

- The chart provides a clear pattern on how speed and the number of vehicles change as day moves systematically up in a week. The seven points are connected by a straight line for both the speed and vehicle trend lines.
- The vertical scales are significantly different. Speed uses the left vertical axis. The number of vehicles uses the right axis.
- The chart is journal quality and can be used directly for publication.

Case 7B: SAS Codes (E72c7b)

```
GOPTIONS RESET=GLOBAL;
TITLE FONT=Swissb HEIGHT=16PT COLOR=BLACK 'Overlay AM Speed and
Vehicle Numbers'  ;
FOOTNOTE1 Font=oldeng HEIGHT=12PT JUSTIFY=LEFT 'Jan 31, 2019';

SYMBOL1 VALUE=SQUARE COLOR=BLACK HEIGHT=10PT I=join L= 1;
SYMBOL2 VALUE=Triangle COLOR=BLACK HEIGHT=10PT I=join L=2;

AXIS1 ORDER=('Sunday' 'Monday' 'Tuesday' 'Wednesday' 'Thursday'
'Friday' 'Saturday' )
        VALUE=(FONT=swiss Color=black HEIGHT=10PT )
```

```
         LABEL = (HEIGHT=1.5 FONT=swissb COLOR=black 'Day of
Week')
        OFFSET=(10PT,10PT)
              WIDTH=4;

AXIS2 ORDER =(0 TO 80 BY 10)
        VALUE=(FONT=swiss HEIGHT=10PT color=black )
        MAJOR= (HEIGHT=18PT)
        MINOR= (N=4 HEIGHT=8PT)
        LABEL=(HEIGHT=15pt FONT=swissb ANGLE=90 JUSTIFY=C 'Mean
Speed (mph)');

AXIS3 ORDER =(1500 TO 3200 BY 300)
        VALUE=(FONT=swiss Color=black HEIGHT=12PT )
        MAJOR= (HEIGHT=18PT)
        MINOR= (N=4 HEIGHT=8PT)
        LABEL=(FONT=swissb HEIGHT=15pt ANGLE=90 JUSTIFY=C 'Number
of Vehicles');

LEGEND1         POSITION=(top left inside);
LEGEND2         POSITION=(top right inside);

LIBNAME mydata 'c:\sas_exercise';

DATA mydata.as2;
set mydata.average_speed;

RUN;

DATA mydata.as2;
set mydata.average_speed;
IF DOW='Sunday' THEN DOWID=1 ;
IF DOW ='Monday' THEN DOWID=2 ;
IF DOW='Tuesday' THEN DOWID=3;
IF DOW ='Wednesday' THEN DOWID=4;
IF DOW='Thursday' THEN DOWID=5;
IF DOW ='Friday' THEN DOWID=6;
if DOW='Saturday' THEN DOWID=7;
RUN;

PROC Sort data=mydata.as2;
BY DOWID;
RUN;


RUN;
```

```
PROC GPLOT DATA=mydata.as2;
PLOT AMSpeed*DOW /HAXIS=axis1 VAXIS=axis2 LEGEND=Legend1
AUTOVREF;
PLOT2 AMVehicle*DOW/HAXIS =axis1 VAXIS=axis3 LEGEND=legend2
AUTOVREF;
RUN;
QUIT;
```

GOPTIONS is the command to declare graphical options. RESET=GLOBAL instructs SAS to reset all graphical options to default.

The TITLE lists title-related specifications:

FONT=Swissb (font type)
HEIGHT=16PT (title font size)
COLOR=BLACK (font color)

"Overlay .... Numbers" (actual title).

The FOOTNOTE1 lists all footnote-related specifications:

Font=oldeng (font type)
HEIGHT=12PT (font size)
JUSTIFY=LEFT (footnote location justification, left, right, or center)
"Jan 31, 2019" (actual footnote).

SYMBOL1 specifies that the first symbol (if needed by any SAS procedure) is a square (VALUE=SQUARE), having the color of black (COLOR=BLACK), size of 10 pt (HEIGHT=10PT), simple joint (I=join), and line style of 1 (L=1).

SYMBOL2 specifies that the second symbol (if needed by any SAS procedure) is a triangle (VALUE=Triangle), having the color of black (COLOR=BLACK), size of 10 pt (HEIGHT=10PT), simple joint (I=join), and line style of 2 (L=2).

AXIS1 lists its specifications as follows:

ORDER=('Sunday'.... 'Saturday') arranges the sequence of AXIS1 values to be displayed from left to right. It starts with Sunday, then Monday ... and end with Saturday in a chronological order.

VALUE=(FONT=swiss Color=black HEIGHT=10PT) tells SAS that values to be displayed along the AXIS1 should use the 10 pt black swiss font.

LABEL=(HEIGHT=1.5 FONT=swissb COLOR=black) defines that "Day of Week" is the AXIS1 title and it is to be displayed with the 1.5 units black swissb font.

OFFSET=(10PT,10PT) tells SAS to shift the AXIS1 starting point 10 points up and 10 points right from the default location.

WIDTH=4 specifies that the plot frame width has a factor of 4.

AXIS2 lists its specifications as follows:

ORDER=(0 TO 80 BY 10) instructs to start AXIS2 at 0 and end at 80 with an increment of 10.

VALUE=(FONT=swiss HEIGHT=10PT color=black) tells SAS to use black 10 pt swiss font to display all the numbers along AXIS2.

MAJOR=(HEIGHT=18PT) instructs SAS to have the major tick mark with a height of 18 points.

MINOR=(N=4 HEIGHT=8PT) instructs SAS to insert 4 minor tick marks between each pair of major tick marks with a height of 8 points.
LABEL=(HEIGHT=15pt, FONT=swissb ANGLE=90 JUSTIFY=C) defines AXIS2 title to be displayed with a 15 pt swissb font, rotated clockwise 90 degrees from its normal position, and center-justified. "Mean Speed (mph)" is the actual AXIS2 title.
AXIS3 lists its specifications:
ORDER=(1500 TO 3200 BY 300) instructs SAS to start AXIS3 at 1500 and end at 3200 with an increment of 300.
VALUE=(FONT=swiss Color=black HEIGHT=12PT) tells SAS to use the black 12 pt swiss font to display the AXIS3 numbers.
MAJOR=(HEIGHT=18PT) instructs SAS to have the major tick mark with a height of 18 points.
MINOR=(N=4 HEIGHT=8PT) instructs SAS to insert 4 minor tick marks between each pair of major tick marks with a height of 8 points.
LABEL=(FONT=swissb HEIGHT=15pt ANGLE=90 JUSTIFY=C) defines AXIS3 title "Number of Vehicles" is to be centrally displayed and rotated clockwise 90 degrees from its normal position with the 15 pt swissb font.

LEGNED1 POSITION= (top left inside) directs SAS to put the legend (plot of AXIS3) for AMspeed at the top left and within the plot frame.

LEGNED2 POSITION= (top right inside) directs SAS to put the legend (plot of AXIS3) for AMvehicles at the top right and within the plot frame.

LIBNAME defines a library called mydata representing the folder of "c:\sas_exercise."

DATA mydata.as2 defines a library named mydata representing the folder of c:\sas_exercise.

set mydata.average_speed tells SAS to read a datafile named average_speed.sas7bdat in the mydata folder.

IF DOW="Sunday" THEN DOWID=1 creates a new variable called DOWID and assigns a value of 1 if the condition is met.

IF DOW ... THEN ... fills the observations of all DOWID, respectively, to the corresponding DOW value.

PROC Sort data=mydata.as2 invokes the sort procedure for data as2 in the mydata library. Sorting is done by DOWID.

PROC GPLOT invokes the GPLOT procedure for the dataset as2.sas7bdat in the mydata library.

*PLOT AMSpeed*DOW/ HAXIS=axis1 VAXIS=axis2 LEGEND=legend1 instructs SAS to plot the AMSpeed against DOW. It further specifies that the horizontal axis is axis1 and the vertical axis is axis2, and the plot uses the legend of legend1.*

*PLOT2 AMVehicle*DOW/ HAXIS=axis1 VAXIS=axis2 LEGEND=legend2 instructs SAS to plot the AMVehicle against DOW. It further specifies that the horizontal axis is axis1 and the vertical axis is axis2, and the plot uses the legend of legend2.*

AUTOVREF invokes the plotting of vertical (V) reference (REF) lines at every major tick marks.

RUN instructs SAS to execute the PROC GPLOT procedure.

QUIT terminates the procedure.

7.3 PROC GCHART

The **PROC GCHART** procedure can produce several chart types, including pie charts, vertical histograms (hbar), horizontal histograms (vbar), vertical bars, and block diagrams. The **PROC GCHART** has data summary capabilities covering 1) frequency, 2) average, and 3) sum.

The utilization of the PROC GCHART is listed below by using the same data as used in Section 7.2 cases 1–5.

Example: 7.3

Data: myspeed_plot.sas7bdat

The speedplot data have eight variables where speeds at two different segments (**SegID**) of I95 (**Road**) are measured throughout two-week periods (**DOW**) for both the morning (AM) and afternoon (PM) peak commuting time (**Time**). A total of 28 observations (28 rows) are recorded. In addition, the numbers of vehicles (**Vehicles**) observed for having the measured speed are also recorded. The full data is listed below.

Road	SegID	SegLength	DOW	Time	WKWD	Speed	Vehicles
I95	PG101	2.4	Monday	AM	weekday	35	3218
I95	PG101	2.4	Tuesday	AM	weekday	42	2690
I95	PG101	2.4	Wednesday	AM	weekday	32	3111
I95	PG101	2.4	Thursday	AM	weekday	58	2631
I95	PG101	2.4	Friday	AM	weekday	52	2860
I95	PG101	2.4	Saturday	AM	weekend	74	1989
I95	PG101	2.4	Sunday	AM	weekend	60	2421
I95	PG101	2.4	Monday	PM	weekday	36	3098
I95	PG101	2.4	Tuesday	PM	weekday	40	2987
I95	PG101	2.4	Wednesday	PM	weekday	28	3981
I95	PG101	2.4	Thursday	PM	weekday	36	3216
I95	PG101	2.4	Friday	PM	weekday	70	2122
I95	PG101	2.4	Saturday	PM	weekend	71	1898
I95	PG101	2.4	Sunday	PM	weekend	56	2122
I95	PG515	0.62	Monday	AM	weekday	39	3099
I95	PG515	0.62	Tuesday	AM	weekday	38	2591
I95	PG515	0.62	Wednesday	AM	weekday	36	2996
I95	PG515	0.62	Thursday	AM	weekday	24	3900
I95	PG515	0.62	Friday	AM	weekday	66	2754
I95	PG515	0.62	Saturday	AM	weekend	68	1916
I95	PG515	0.62	Sunday	AM	weekend	45	2332
I95	PG515	0.62	Monday	PM	weekday	41	2984
I95	PG515	0.62	Tuesday	PM	weekday	45	2877
I95	PG515	0.62	Wednesday	PM	weekday	25	3834
I95	PG515	0.62	Thursday	PM	weekday	52	2400
I95	PG515	0.62	Friday	PM	weekday	54	2044
I95	PG515	0.62	Saturday	PM	weekend	64	1828
I95	PG515	0.62	Sunday	PM	weekend	62	2044

Case 1: Speed Frequency Plot by Default

Case 1: Result

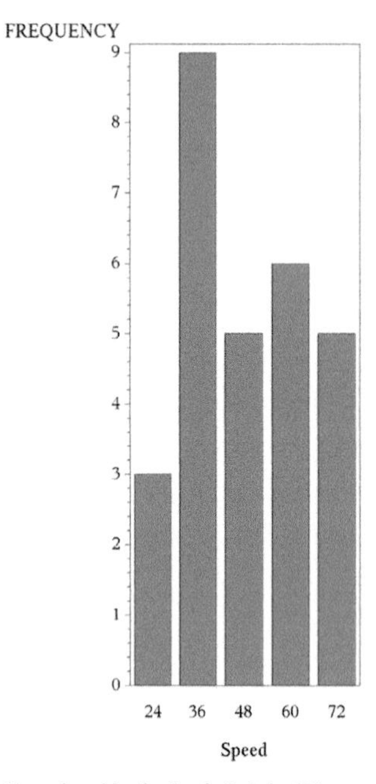

- The frequency plot shows the observed speed distribution pattern by first dividing speed data into different bins. For each bin, the number of observations is counted (# of occurrence) and plotted on the Y-axis.

Case 1: SAS Codes (E73c1)

```
TITLE COLOR=BLACK FONT=SWISSB
J=C 'Frequency Plot - Default';
LIBNAME MYDATA
'C:\SAS_EXERCISE';
PROC GCHART DATA=mydata.as3;
VBAR SPEED/ TYPE= Frequency ;
RUN;
QUIT;
```

The TITLE lists title-related specifications:
COLOR=BLACK (font color)
FONT=SWISSB (font type)
J=C (Justification= center, left, or right)
"Frequency …." (actual title).
LIBNAME defines a library called mydata representing the folder of "C:\SAS_EXERCISE."
PROC GCHART DATA=mydata.as3 invokes SAS's GCHART function for the dataset named as3.sas7bdat in the mydata library.
VBAR declares that it is a vertical bar chart for the variable SPEED showing Frequency distribution (TYPE=Frequency).
RUN instructs SAS to execute PROC GCHART the procedure.
QUIT terminates the procedure.

Case 2: Speed Frequency Plot with Discrete Specification

The "Discrete" specification treats each observed speed as a separate bin.

Case 2: Result

- The "discrete" specification treats each unique speed as a separate bin. Bins with more than 1 observation indicate the presence of multiple observations with the same speed.

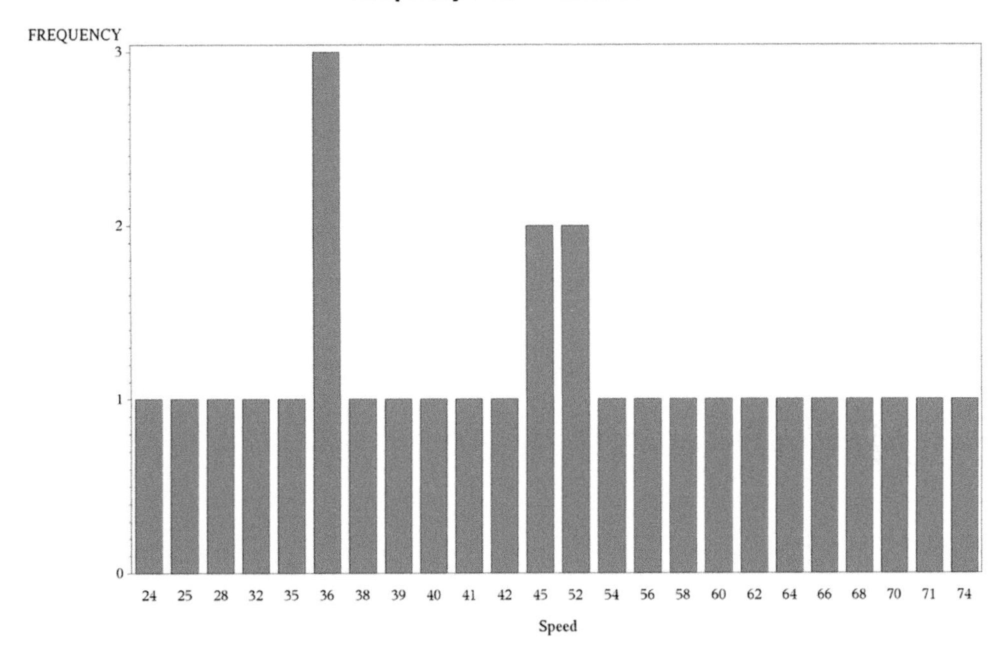

Completed in the Study Lab by BJ

Case 2: SAS Codes (E73c2)

```
TITLE COLOR=BLACK
FONT=SWISSB J=C
'Frequency Plot
- Discrete';
LIBNAME MYDATA 'C:\
SAS_EXERCISE';
PROC GCHART
DATA=mydata.as3;
VBAR SPEED/ TYPE=
Frequency DISCRETE;
RUN;
QUIT;
```

The TITLE lists title-related specifications:
COLOR=BLACK (font color)
FONT=SWISSB (font type)
J=C (Justification= center, left, or right)
"Frequency …Discrete" (actual title).

LIBNAME defines a library called mydata representing the folder of "C:\SAS_EXERCISE."

PROC GCHART DATA=mydata.as3 invokes SAS's GCHART function for the dataset named as3.sas7bdat in the mydata library.

VBAR declares that it is a vertical bar chart for the variable SPEED showing Frequency distribution (TYPE=Frequency).

The DISCRETE specification asks SAS to not aggregate speeds into groups. Instead, every unique speed is a group.

RUN instructs SAS to execute the PROC GCHART procedure.

QUIT terminates the procedure.

Case 3: Speed–Frequency Plot with Predefined # of Bars

A user can specify how many bars (bins) the user wants to use in which to divide the speed. Here, we would like to have six bars (bins).

Case 3: Result

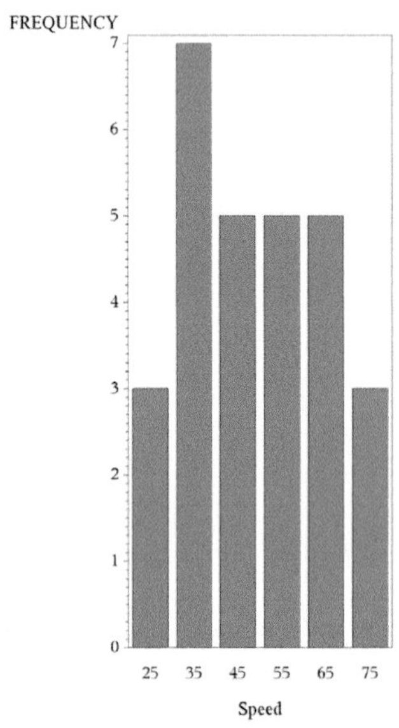

- The user specifies 6 bins for the speed.

Case 3: SAS Codes (E73c3)

```
TITLE COLOR=BLACK FONT=SWISSB J=C 'Frequency Plot - 6 Levels';
LIBNAME MYDATA 'C:\SAS_EXERCISE';
PROC GCHART DATA=mydata.as3;
VBAR SPEED/ TYPE= Frequency LEVELS=6;
RUN;
QUIT;
```

The TITLE lists title-related specifications:
COLOR=BLACK (font color)
FONT=SWISSB (font type)
J=C (Justification= center, left, or right)
"Frequency Plot -6 Levels" (actual title).
LIBNAME defines a library called mydata representing the folder of "C:\SAS_EXERCISE."
PROC GCHART DATA=mydata.as3 invokes SAS's GCHART function for the dataset named as3.sas7bdat in the mydata library.

VBAR declares that it is a vertical bar chart for the variable SPEED showing Frequency distribution (TYPE=Frequency).

The LEVELS=6 specification asks SAS to aggregate speeds into 6 groups and counts the number of observations in each group.

RUN instructs SAS to execute the PROC GCHART procedure.

QUIT terminates the procedure.

Case 4: Speed Frequency by Time with Pattern Options

Case 4: Result

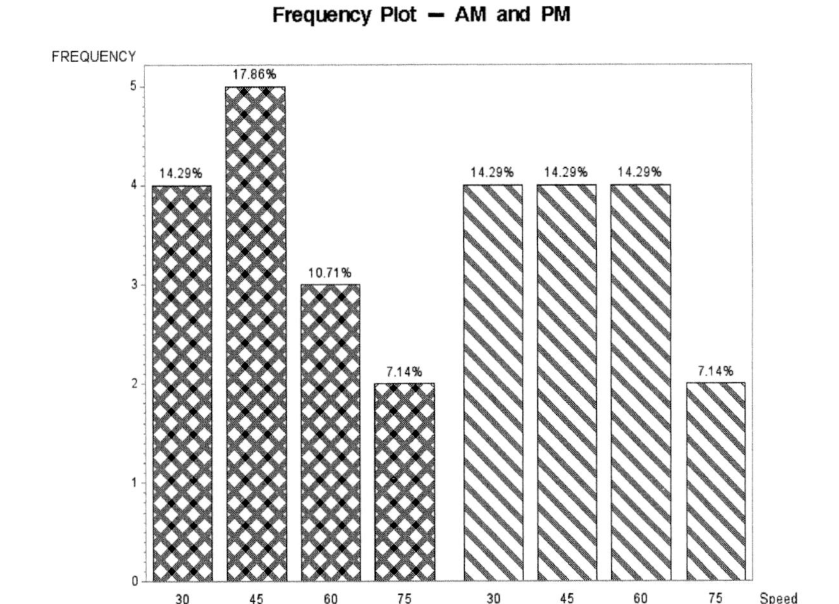

Case 4: SAS Codes (E73c4)

```
GOPTION RESET=Global
RESET=PATTERN CPATTERN=BLACK;

TITLE COLOR=BLACK FONT=SWISSB J=C 'Frequency Plot - AM and PM';
PATTERN1 VALUE=X5  ;
PATTERN2 VALUE=L5 ;

LIBNAME MYDATA 'C:\SAS_EXERCISE';
PROC GCHART DATA=mydata.as3;
VBAR SPEED/ TYPE=Freq LEVELS=4 Width =6 OUTSIDE=percent GROUP=TIME
PATTERNID=GROUP;
RUN;
QUIT;
```

GOPTION is the command to declare graphical options. RESET=Global instructs SAS to reset all graphical options to default. In addition, RESET=Pattern reset all patterns to default CPATTERN=BLACK sets pattern color as black.

The TITLE lists title-related specifications:

COLOR=BLACK (font color)

FONT=SWISSB (font type)

J=C (Justification= center, left, or right)

"Frequency Plot - AMand PM" (actual title).

PATTERN1 VALUE=X5 specifies that the PATTERN1 is cross-hatched lines with level 5 shade.

PATTERN2 VALUE=L5 specifies that the PATTERN2 is left-slanted hatching with level 5 shade.

LIBNAME defines a library called mydata representing the folder of "C:\SAS_EXERCISE."

PROC GCHART DATA=mydata.as3 invokes SAS's GCHART function for the dataset named as3.sas7bdat in the mydata library.

VBAR declares that it is a vertical bar chart for the variable SPEED showing Frequency distribution (TYPE=Freq).

The LEVELS=4 specification asks SAS to aggregate speeds into 4 groups and count the number of observations in each group.

Width=6 specifies the width of the vertical bar (you need to test it out with different scales to find the optimal arrangement).

OUTSIDE=percent tells SAS to label the vertical bar with the percent value and put the label outside the vertical bar area.

GROUP=TIME instructs SAS to group the speed by TIME where the AMSpeed has 4 levels and PMSpeed has 4 levels and display them separately .

PATTERNID=GROUP tells SAS to match the pattern based on matching pATTERNID with the group number (e.g., GROUP=TIME=AM=PATTERN1, GROUP=TIME=PM=PATTERN2)

RUN instructs SAS to execute the PROC GCHART procedure.

QUIT terminates the procedure.

Case 5: Mean Speed by DOW – Default Order (Automatically Alphabetically Arranged)

Case 5: Result

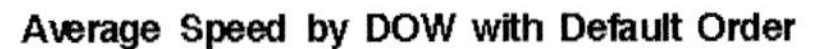

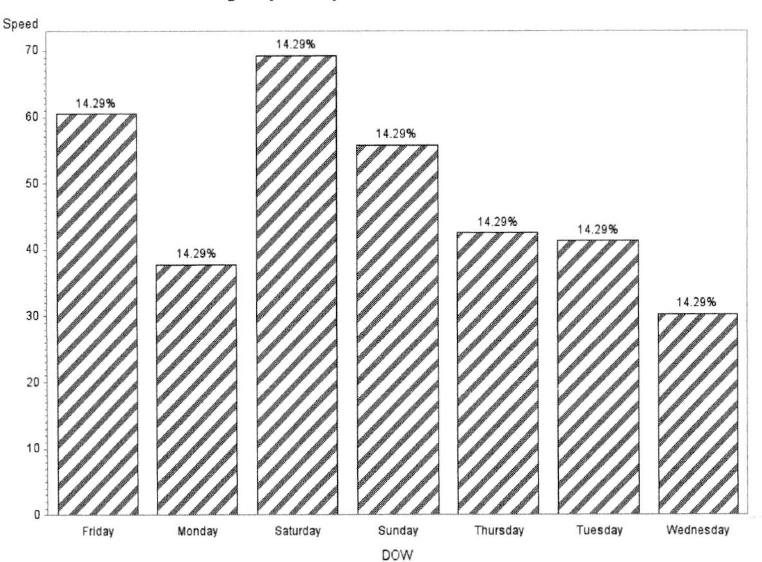

Case 5: SAS Codes (E73c5)

```
GOPTION RESET=Global ;
Pattern value=X5;
TITLE COLOR=BLACK FONT=SWISSB J=C 'Average Speed by DOW with
Default Order';
LIBNAME MYDATA 'C:\SAS_EXERCISE';
PROC GCHART DATA=mydata.as3;
VBAR DOW/TYPE=MEAN SUMVAR=SPEED OUTSIDE=percent ;
RUN;
QUIT;
```

GOPTION is the command to declare other options. RESET=Global instructs SAS to reset all graphical options to default.

Pattern value=X5 tells SAS to use cross-hatched line with level 5 shade for the pattern.

The TITLE lists title-related specifications:

COLOR=BLACK (font color)

FONT=SWISSB (font type)

J=C (Justification= center, left, or right)

"Average Speed by DOW with Default Order" (actual title).

LIBNAME defines a library called mydata representing the folder of "C:\SAS_EXERCISE."

PROC GCHART DATA=mydata.as3 invokes SAS's GCHART function for the dataset named as3.sas7bdat in the mydata library.

VBAR declares that it is a vertical bar chart where the variable DOW is used for the horizontal axis and the vertical axis is the MEAN (TYPE=MEAN) for the variable SPEED (SUMVAR=SPEED).

OUTSIDE=percent tells SAS to label the vertical bar with the percent value and put the label outside the vertical bar area.

RUN instructs SAS to execute the PROC GCHART procedure.

QUIT terminates the procedure.

Case 6: Mean Speed by DOW with Options

Case 6: Result

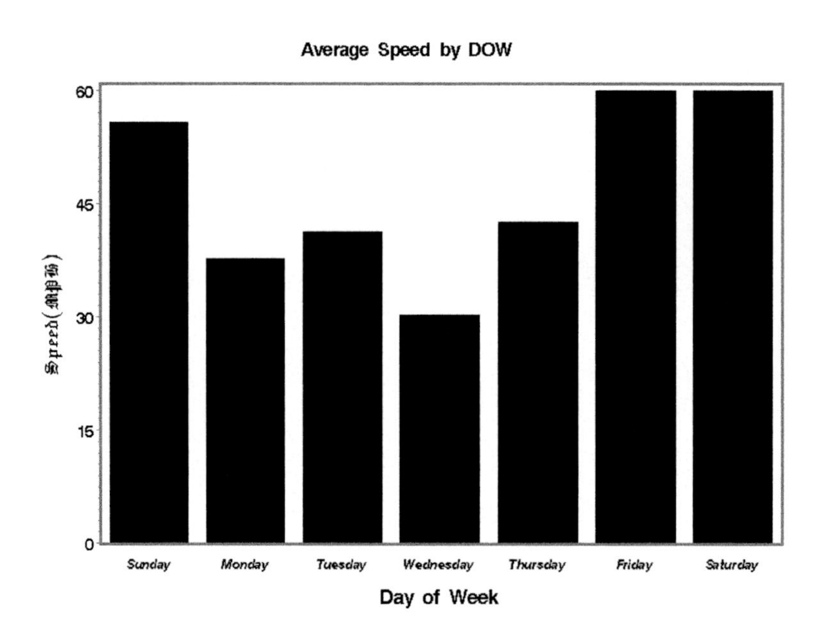

Case 6: SAS Codes (E73c6)

```
GOPTION RESET=Global CPATTERN=black;
TITLE COLOR=BLACK FONT=SWISSB J=C 'Average Speed by DOW';
PATTERN1 VALUE=s;
AXIS1 Order=(0 to 70 by 15)
      Value =(Font=swissb color=black height =12 pt)
      LABEL=(Font = Oldeng A=90 H=16pt 'Speed(MPH)');
AXIS2 ORDER=('Sunday' 'Monday' 'Tuesday' 'Wednesday' 'Thursday'
'Friday' 'Saturday')
      VALUE=(Font=swissbi Color=black HEIGHT=10PT )
        LABEL = (HEIGHT=15pt FONT=swissb COLOR=black 'Day of
Week')
        WIDTH=2;
LIBNAME MYDATA 'C:\SAS_EXERCISE';
```

```
PROC GCHART DATA=mydata.as3;
VBAR DOW/ TYPE=MEAN SUMVAR=SPEED WIDTH=12 RAXIS=axis1 MAXIS=axis2;
RUN;
QUIT;
```

GOPTION is the command to declare graphical options. RESET=Global instructs SAS to reset all graphical options to default. CPATTERN=black tells SAS to use black as the pattern color.

The TITLE lists title-related specifications:

COLOR=BLACK (font color)

FONT=SWISSB (font type)

J=C (Justification= center, left, or right)

"Average Speed by DOW" (actual title).

PATTERN1 VALUE=s specifies that the PATTERN1 is solid.

AXIS1 lists its specifications:

Order=(0 to 70 by 15) starts the AXIS1 at 0 and end at 70 with an increment of 15.

Value =(Font=swissb color =black height=12 pt) instructs SAS to use the black 12 pt swissb font for AXIS1 values.

LABEL=(Font=Oldeng A=90 H=16pt "Speed (MPH)" tells SAS that the label "Speed (MPH)" should be rotated 90 degrees clockwise before placement and displayed with 16 pt with Oldeng font.

AXIS2 lists its specifications:

ORDER=('Sunday'.... 'Saturday') arranges the sequence of AXIS2 values to be displayed from left to right. It starts with Sunday, then Monday ... and ends with Saturday.

VALUE=(Font=swissbi Color=black HEIGHT=10PT) tells SAS that values to be displayed along the AXIS2 should use the black 10 pt swissbi font.

LABEL=(HEIGHT=15pt FONT=swissb COLOR=black) defines the title "Day of Week" for AXIS2 will use the black 15 pt swissb font.

WIDTH=2 specifies that the plot frame width has a factor of 2.

LIBNAME defines a library called mydata representing the folder of "C:\SAS_EXERCISE."

PROC GCHART DATA=mydata.as3 invokes SAS's GCHART function for the dataset named as3.sas7bdat in the mydata library.

VBAR declares that it is a vertical bar chart where the variable DOW is used for the horizontal axis and the vertical axis is MEAN (TYPE=MEAN) for the variable SPEED (SUMVAR=SPEED). WIDTH=12 sets the bar width as 12. RAXIS=axis1 and MAXIS =axis2; tells SAS that the Response Axis (vertical axis) is axis 1 and the midpoint axis is the axis 2

RUN instructs SAS to execute the PROC GCHART procedure.

QUIT terminates the procedure.

Case 7: Mean Speed by DOW and Time with Options

Case 7: Result

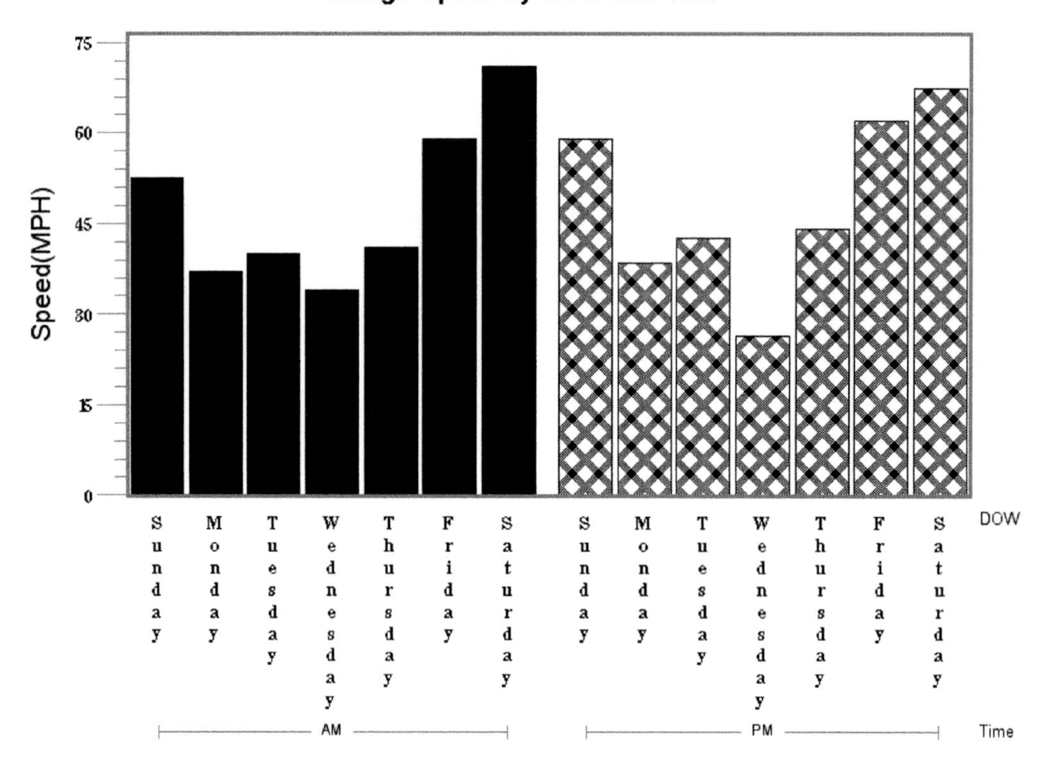

Case 7: SAS Codes (E73c7)

```
GOPTION RESET=Global
CPATTERN=black;
TITLE COLOR=BLACK FONT=SWISSB J=C 'Average Speed by DOW and Time';

PATTERN1 VALUE=S;
PATTERN2 VALUE=x5;

AXIS1 Order =(0 to 80 by 15)
        MAJOR= (HEIGHT=18PT)
        MINOR= (N=4 HEIGHT=8PT)
        VALUE=(Font=centb HEIGHT=10PT Color=black)
      LABEL=(A=90 H=16pt FONT=Oldeng 'Speed(MPH)');

AXIS2 ORDER=('Sunday' 'Monday' 'Tuesday' 'Wednesday' 'Thursday'
'Friday' 'Saturday')
      VALUE=(Font=centb HEIGHT=10PT Color=black)
      WIDTH=2;

LIBNAME MYDATA 'C:\SAS_EXERCISE';
PROC GCHART DATA=mydata.as3;
```

```
VBAR DOW/ TYPE=MEAN SUMVAR=SPEED WIDTH=6 RAXIS=axis1 MAXIS=axis2
GROUP=TIME PATTERNID=GROUP;
RUN;
QUIT;
```

GOPTION is the command to declare graphical options. RESET=Global instructs SAS to reset all graphical options to default. CPATTERN=black tells SAS to use black as the pattern color.

The TITLE lists title-related specifications:

COLOR=BLACK (font color)
FONT=SWISSB (font type)
J=C (Justification= center, left, or right)
"Average Speed by DOW and Time" (actual title).

PATTERN1 VALUE=S specifies that the PATTERN1 is solid.

PATTERN2 VALUE =x5 specifies that the PATTERN2 is cross-hatched lines with level of 5 shade.

AXIS1 lists all its specifications as follows:

Order=(0 to 80 by 15) instructs SAS to start AXIS1 at 0 and end at 80 with an increment of 15.

MAJOR= (HEIGHT=18PT) specifies that the major tick mark is 18 pt. MINOR= (N=4 HEIGHT=8PT) tells SAS that 4 minor ticks should be inserted between a pair of major tick marks with a height of 8 pt.

VALUE=(Font=centb HEIGHT=10PT Color =black) tells SAS to use black 10 pt centb font to display all the AXIS1 values.

LABEL=(A=90 H=16pt FONT=Oldeng 'Speed (MPH)" tells SAS that the AXIS1 title "Speed (MPH)" should be rotated 90 degrees clockwise before placement and should be centrally displayed with the 16pt Oldeng font.

AXIS2 lists all its specifications as follows:

ORDER=('Sunday'.... 'Saturday') arranges the sequence of AXIS2 values to be displayed from left to right. It starts with Sunday, then Monday … and ends with Saturday.

VALUE=(Font=centb HEIGHT=10PT Color=black) tells SAS that values to be displayed along AXIS2 should use the black 10 pt centb font.

WIDTH=2 specifies that the plot frame width has a factor of 2.

LIBNAME defines a library called mydata representing the folder of "C:\SAS_EXERCISE."

PROC GCHART DATA=mydata.as3 invokes SAS's GCHART function for the dataset named as3.sas7bdat in the mydata library.

VBAR declares that it is a vertical bar chart where the variable DOW is used for the horizontal axis and the vertical axis is MEAN (TYPE=MEAN) for the variable SPEED (SUMVAR=SPEED). WIDTH=6 sets the bar width as 6. RAXIS=axis1 and MAXIS=axis2 tell SAS axis1 is the response axis (vertical axis) and axis2 is the midpoint axis

GROUP=TIME tells SAS that AM and PM analysis should be separate. Patterns used to identify the AM and PM (PATTERNID) should follow the GROUP IDS.

RUN instructs SAS to execute the PROC GCHART procedure.

QUIT terminates the procedure.

Case 8: Mean Speed by DOW and Time with Paired Bars

Case 8: Result

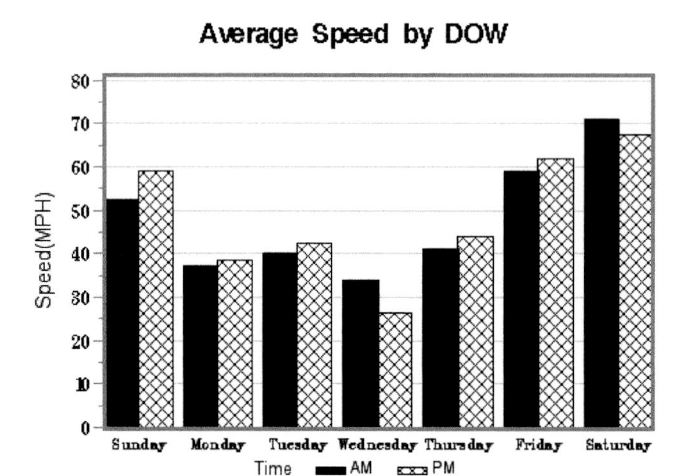

Case 8: SAS Codes (E73c8)

```
GOPTIONS RESET=all
HSIZE=5INCHES VSIZE=3.5INCHES;

TITLE COLOR=BLACK FONT=SWISSB J=C 'Average Speed by DOW';

PATTERN1 VALUE=S C=BLACK;
PATTERN2 VALUE=x1 C=BLACK;

AXIS1 order=(0 to 80 by 10)
      major= (Height=6pt)
      MINOR=(N=1 Height=3pt)
     VALUE=(Font=centb HEIGHT=10PT Color=black)
      LABEL=(A=90 H=11pt FONT=centb 'Speed(MPH)');

AXIS2 value=none
      LABEL=none;

AXIS3 ORDER=('Sunday' 'Monday' 'Tuesday' 'Wednesday' 'Thursday'
'Friday' 'Saturday')
      VALUE=(HEIGHT=8PT FONT=Triplex)
        LABEL=None
        WIDTH=2;

LIBNAME MYDATA 'C:\SAS_EXERCISE';
```

```
PROC GCHART DATA=mydata.as3;
VBAR TIME/ TYPE=MEAN SUMVAR=SPEED RAXIS=axis1 MAXIS=axis2
GAXIS=axis3 GROUP=DOW SUBGROUP=TIME WIDTH=6 SPACE=0 GSPACE=2
AUTOREF CLIPREF CREF=GRAYDD;
RUN;
QUIT;
```

GOPTIONS is the command to declare graphical options. RESET=all instructs SAS to reset all SAS graphical options to default.

HSIZE=5INCHES VSIZE=3.5INCHES tells SAS that the output for the current procedure is 5 inches by 3.5 inches.

The TITLE lists title-related specifications:

COLOR=BLACK (font color)

FONT=SWISSB (font type)

J=C (Justification= center, left, or right)

"Average Speed by DOW" (actual title).

PATTERN1 VALUE=S C=BLACK specifies that the PATTERN1 is solid with a color of black.

PATTERN2 VALUE =x1 C=BLACK specifies that the PATTERN2 is the cross-hatched line with level 1 shade.

AXIS1 lists its specifications:

order=(0 to 80 by 10) tells SAS to start AXIS1 from 0 and end at 80 with an increment of 10. major= (Height=6pt) specifies that the major tick mark is 6 pt. MINOR=(N=1 Height=3 pt) tells SAS that 1 minor tick mark (N=1) should be inserted between a pair of major tick marks with the height of 3 pt.

VALUE=(Font=centb HEIGHT=10PT Color=black) tells SAS that values to be displayed along the AXIS1 is 10 pt centb font in black.

LABEL=(A=90 H=11pt FONT=centb "Speed(MPH)" tells SAS that the title should be rotated 90 degrees clockwise and use the 11pt centb font. "Speed(MPH)" is the actual AXIS1 title.

AXIS2 value=none LABEL=none tells SAS that there is no title for AXIS2 and no unique specifications for AXIS2.

AXIS3 lists its specifications as follows:

ORDER=('Sunday'.... 'Saturday') arranges the sequence of AXIS3 values to be displayed from left to right. It starts with Sunday, then Monday … and ends with Saturday.

VALUE=(HEIGHT=8PT FONT=Triplex) tells SAS that values to be displayed along AXIS3 is the 8 pt triplex font.

LABEL=None tells SAS that AXIS3 has no axis title.

WIDTH=2 specifies that the plot frame width has a factor of 2.

LIBNAME defines a library called mydata representing the folder of "C:\SAS_EXERCISE."

PROC GCHART DATA=mydata.as3 invokes SAS's GCHART function for the dataset named as3.sas7bdat in the mydata library.

VBAR declares that it is a vertical bar chart where the variable DOW is used for the horizontal axis, and the vertical axis is MEAN (TYPE=MEAN) for the variable SPEED (SUMVAR=SPEED).

RAXIS=axis1 MAXIS =axis2 GAXIS=axis3 defines the response axis (vertical axis) is axis1, midpoint axis is axis2, and the group axis is axis3.
GROUP=DOW tells SAS that analysis should be separated by DOW.
SUBGROUP=TIME instructs SAS to analyze each DOW by TIME (AM and PM).
WIDTH=6 tells SAS each vertical bar has a width of 6. There should be no space (SPACE=0) between AM and PM bars for a given DOW.
GSPACE=2 tells SAS to keep the vertical bar space between each DOW with a unit of 2.
AUTOREF draws reference lines at all major tick marks on the response axis (RAXIS).
CLIPREF clips reference lines at the bars making reference lines appearing as behind bars.
CREF=GRAYDD tells SAS that the color of the reference line is GRAYDD.
RUN instructs SAS to execute the PROC GCHART procedure.
QUIT terminates the procedure.

Case 9: Block Diagram – Default

Case 9: Result

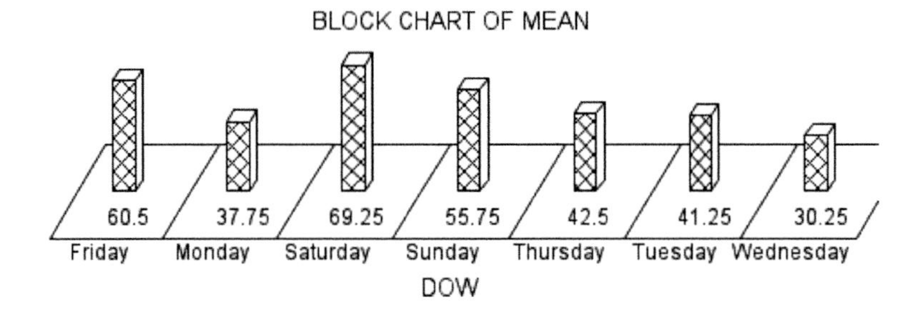

Case 9: SAS Codes (E73c9)

```
GOPTION RESET=Global CPATTERN=black;
TITLE COLOR=BLACK FONT=SWISSB J=C 'Average Speed by Day of Week';
LIBNAME MYDATA 'C:\SAS_EXERCISE';

PROC GCHART DATA=mydata.as3;
BLOCK DOW/ TYPE=MEAN SUMVAR=SPEED ;
RUN;
QUIT;
```

GOPTION is the command to declare graphical options. RESET=Global instructs SAS to reset all SAS graphical options to default. CPATTERN=black tells SAS to use black color for pattern filling.
The TITLE lists title-related specifications:
COLOR=BLACK (font color)

FONT=SWISSB (font type)
J=C (Justification= center, left, or right)
"Average Speed by Day of Week" (actual title).
LIBNAME defines a library called mydata representing the folder of "C:\SAS_EXERCISE."
PROC GCHART DATA=mydata.as3 invokes SAS's GCHART function for the dataset named as3.sas7bdat in the mydata library.
BLOCK declares that the block chart procedure is to be used. The variable DOW is used for the midpoint axis (MAXIS) and the vertical axis is average (TYPE=MEAN) for the variable SPEED (SUMVAR=SPEED).
RUN instructs SAS to execute the PROC GCHART procedure.
QUIT terminates the procedure.

Case 10: Block Diagram

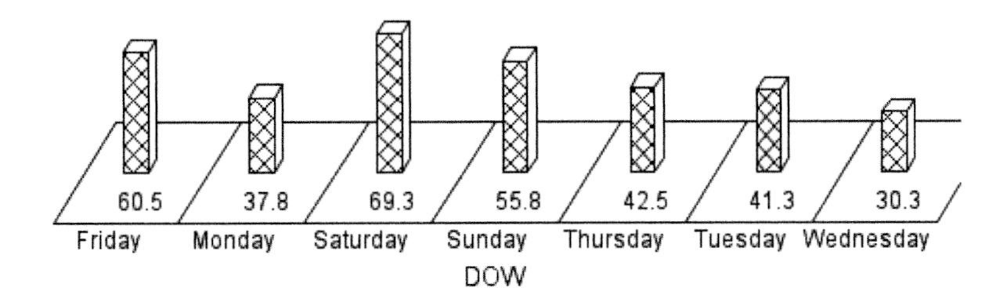

Case 10: SAS Codes (E73c10)

```
GOPTION RESET=Global CPATTERN=black;
TITLE COLOR=BLACK FONT=SWISSB J=C 'Average Speed by Day of Week';
LIBNAME MYDATA 'C:\SAS_EXERCISE';

PROC GCHART DATA=mydata.as3;
FORMAT speed 4.1;
BLOCK DOW/ TYPE=MEAN SUMVAR=SPEED NOHEADING ;
RUN;
QUIT;
```

GOPTION is the command to declare graphical options. RESET=Global instructs SAS to reset all SAS pattern options to default. CPATTERN=black tells SAS to use black color for the pattern filling.
The TITLE lists title-related specifications:
COLOR=BLACK (font color)
FONT=SWISSB (font type)
J=C (Justification= center, left, or right)
"Average Speed by Day of Week" (actual title).
LIBNAME defines a library called mydata representing the folder of "C:\SAS_EXERCISE."

PROC GCHART DATA=mydata.as3 invokes SAS's GCHART function for the dataset named as3.sas7bdat in the mydata library.

FROMAT speed 4.1 requests SAS to display the speed with 1 decimal point.

BLOCK declares that the block chart procedure is to be used. The variable DOW is used for the midpoint axis (MAXIS) and the vertical axis is average (TYPE=MEAN) for the variable SPEED (SUMVAR=SPEED).

The NOHEADING command option removes the default "Block Chart of Mean" from the chart as displayed in the E73c8 Case 9.

RUN instructs SAS to execute the PROC GCHART procedure.

QUIT terminates the procedure.

Case 11: Block Diagram with Options

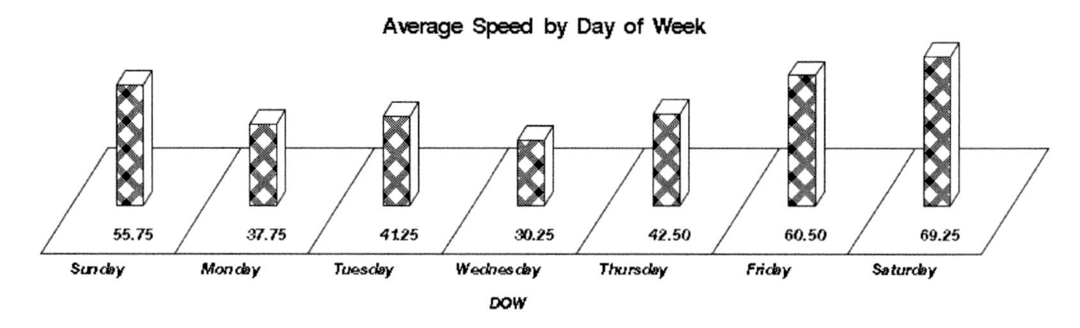

Case 11: SAS Codes (E73c11)

```
GOPTIONS RESET=All
FTEXT=ARIAL HTEXT=9PT ;
PATTERN VALUE=X3 C=BLACK ;
TITLE COLOR=BLACK FONT=SWISSB H=12PT J=C 'Average Speed by Day of
Week';
LIBNAME MYDATA 'C:\SAS_EXERCISE';
PROC GCHART DATA=mydata.as3;
FORMAT SPEED 5.2;

BLOCK DOW / TYPE=MEAN SUMVAR=SPEED
MIDPOINTS= 'Sunday' 'Monday' 'Tuesday' 'Wednesday' 'Thursday'
'Friday' 'Saturday'
BLOCKMAX=70  NOHEADING  ;
RUN;
QUIT;
```

GOPTIONS is the command to declare graphical options. RESET=ALL instructs SAS to reset all SAS graphical options to default.

FTEXT=ARIAL HTEXT=9PT specifies that in the absence of other customized specifications, all labels used use the 9-point Arial font.

PATTERN VALUE=X3 specifies that the pattern is the cross-hatched line with level 3 black shade (C=BLACK).

The TITLE lists title-related specifications:
COLOR=BLACK (font color)
FONT=SWISSB (font type)
H=12PT (font size)
J=C (Justification= center, left, or right)
"Average Speed by Day of Week" (actual title).
LIBNAME defines a library called mydata representing the folder of "C:\SAS_EXERCISE."
FORMAT SPEED5.2 tells SAS to display the SPEED variable with 2 decimals.
PROC GCHART DATA=mydata.as3 invokes SAS's GCHART function for the dataset named as3.sas7bdat in the mydata library.
BLOCK declares that the block chart procedure is to be used. The variable DOW is used for the midpoint axis (MAXIS) and the vertical axis is average (TYPE=MEAN) for the variable SPEED (SUMVAR=SPEED).
The NOHEADING command option removes the default "Block Chart of Mean" from the chart.
MIDPOINTS="Sunday" "Monday" ... Saturday" arranges the order of the observations associated with the midpoint axis from left to right. BLOCKMAX=70 tells SAS that the maximum of average speed with the chart is 70 (If you have more than one chart to make, this specification enables you to keep a similar scale).
RUN instructs SAS to execute the PROC GCHART procedure.
QUIT terminates the procedure.

Case 12: Pie Chart

Case 12: Result

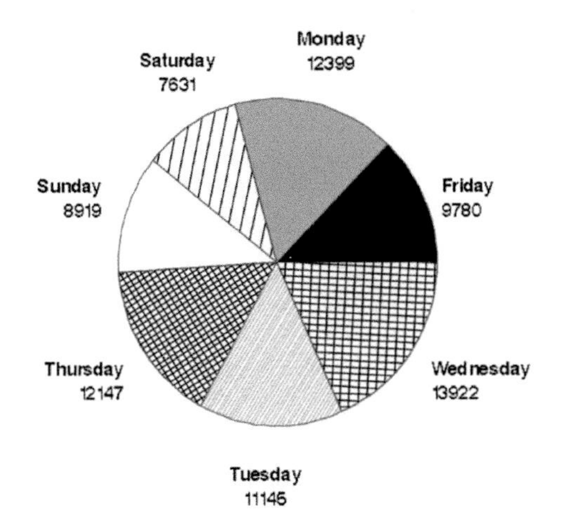

Case 12: SAS Code (E73c12)

```
GOPTIONS RESET=All
HSIZE=5.5INCHES VSIZE=3.5INCHES FTEXT=Swissb HTEXT=8PT;
TITLE1 COLOR=black FONT=swissb HEIGHT=12pt LS=12pt
 "Cumulative Number of Vehicles Observed";
PATTERN1 VALUE=pSOLID COLOR=GRAY00;
PATTERN2 V=pS COLOR=GRAYC0;
PATTERN3 VALUE=P1N45 COLOR=GRAY10;
PATTERN4 V=pEMPTY COLOR=GRAYC0 ;
PATTERN5 VALUE=p4x90 COLOR=GRAY00;
PATTERN6 V=p5N60 COLOR=GRAYC0;
PATTERN7 VALUE=p2x30 COLOR=GRAY00;
LIBNAME mydata 'c:\sas_exercise';
PROC GCHART DATA=MYDATA.as3;
PIE DOW/ TYPE=SUM SUMVAR=vehicles NOHEADER SLICE=OUTSIDE
VALUE=OUTSIDE COUTLINE=GREY;
RUN;
QUIT;
```

GOPTIONS is the command to declare graphical options. RESET=ALL instructs SAS to reset all SAS graphical options to default.

HSIZE=5.5INCHES VSIZE=3.5INCHES tell SAS that the output for the current procedure is 5.5 inches by 3.5 inches. In the absence of any further label specification, font is swissb (FTEXT=Swissb) and height (HTEXT=8PT) is 8 points for all labels.

The TITLE1 lists title-related specifications:

COLOR=black (font color)
FONT=swissb (font type)
HEIGHT=12pt (font size)
LS=12pt (Line spacing above and below the current line)
"Cumulative … Observed" (actual title).

PATTERN1 VALUE=pSOLID COLOR=GRAY00 specifies a Pie (p) chart slice with solid gray00.

PATTERN2 V=pS COLOR=GRAYC0 specifies a Pie (p) chart with solid (S) gray0.

PATTERN3 VALUE=P1N45 – P signifies a Pie chart slice, 1 is shading level, N tells SAS to use parallel lines, and 45 is the angle (degree) of parallel lines.

PATTERN4 V=pEMPTY - p signifies a Pie chart slice, and EMPTY tells SAS to leave filling blank.

PATTERN5 VALUE=p4x90 - p signifies a Pie chart slice, 4 is shading level, x tells SAS to use cross-hatched lines, and 90 is the angle (degree) of the cross-hatched lines.

PATTERN6 V=P5N60 - P signifies a Pie chart slice, 5 is shading level, N tells SAS to use parallel lines, and 60 is the angle (degree) of the parallel lines.

PATTERN7 VALUE=P2x30 - P signifies a Pie chart slice, 2 is shading level, x tells SAS to use cross-hatched lines, and 30 is the angle (degree) of the cross-hatched lines.

LIBNAME defines a library called mydata representing the folder of "c:\sas_exercise."

PROC GCHART DATA=MYDATA.as3 invokes SAS's GCHART function for the dataset named as3.sas7bdat in the mydata library.

PIE declares that the PIE chart procedure is to be used. The variable DOW is used to split the PIE for the variable vehicles (SUMVAR=vehicles). Actual data is the summation of vehicles (TYPE=SUM).

NOHEADER tells SAS not to print the default title of "SUM of Vehicles by DOW."

SLICE=OUTSIDE instructs SAS to label slice on the outside.

VALUE=OUTSIDE tells SAS to label the slice with the actual data on the outside of the slice.

COUTLINE=GREY tells SAS to use gray-colored lines to demark the slice.

RUN instructs SAS to execute the PROC GCHART procedure.

QUIT terminates the procedure.

Case 13: 3D Pie Chart

Case 13: Result

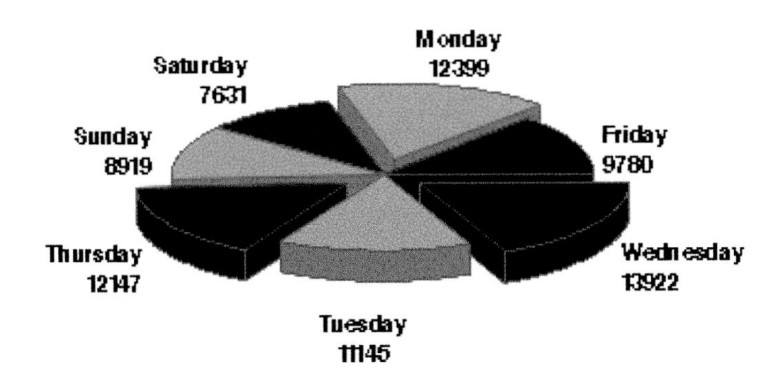

Case 13: SAS Code (E73c13)

```
GOPTIONS RESET=All
HSIZE=5.5INCHES VSIZE=3.5INCHES FTEXT=Swissb HTEXT=8PT;
TITLE1 COLOR=black FONT=swissb HEIGHT=12pt LS=12pt
 "Cumulative Number of Vehicles Observed";
PATTERN1 VALUE=pSOLID COLOR=GRAY00;
PATTERN2 V=pS COLOR=GRAYC0;
PATTERN3 VALUE=P1N45 COLOR=GRAY10;
PATTERN4 V=pEMPTY COLOR=GRAYC0 ;
PATTERN5 VALUE=p4x90 COLOR=GRAY00;
PATTERN6 V=p5N60 COLOR=GRAYC0;
PATTERN7 VALUE=p2x30 COLOR=GRAY00;
LIBNAME mydata 'c:\sas_exercise';

PROC GCHART DATA=MYDATA.as3;
PIE3D DOW/ TYPE=SUM EXPLODE='Monday' 'Thursday' 'Wednesday'
SUMVAR=vehicles NOHEADER SLICE=OUTSIDE VALUE=OUTSIDE COUTLINE=GREY
```

```
;
RUN;
QUIT;
```

GOPTIONS is the command to declare graphical options. RESET=ALL instructs SAS to reset all SAS graphical options to default.

HSIZE=5.5INCHES VSIZE=3.5INCHES tells SAS that the output for the current procedure is 5.5 inches by 3.5 inches. In the absence of any further specification for lables, font is Swissb (FTEXT=Swissb) and height (HTEXT=8PT) is 8 points for all labels.

The TitleTITLE1 lists title -related specifications.:

COLOR=black (font color)

FONT=swissb (font type)

HEIGHT=12pt (font size)

LS=12pt (Line spacing above and below the current line)

"Cumulative ... Observed" (actual title).

PATTERN1 VALUE=pSOLID COLOR=GRAY00 specifies a Pie (p) chart slice with solid gray00.

PATTERN2 V=pS COLOR=GRAYC0 specifies a Pie (p) chart slice with solid (S) gray0.

PATTERN3 VALUE=P1N45 COLOR=GRAY10 – P signifies a Pie chart slice, 1 is the shading level, N tells SAS to use parallel lines, and 45 is the angle (degree) of parallel lines. COLOR=GRAY10 signifies a gray10 color is used for the background.

PATTERN4 V=pEMPTY COLOR=GRAYC0 – p signifies a Pie chart slice, and EMPTY tells SAS to leave filling blank. COLOR=GRAYC0 signifies a grayc0 color is used for the background.

PATTERN5 VALUE=p4x90 COLOR=GRAYC00- p signifies a Pie chart slice, 4 is the shading level, x tells SAS to use cross-hatched lines, and 90 is the angle (degree) of the cross-hatched lines. COLOR=GRAY00 signifies a gray00 color is used for the background.

PATTERN6 V=p5N60 COLOR=GRAYC0 - p signifies a Pie chart slice, 5 is the shading level, N tells SAS to use parallel lines, and 60 is the angle (degree) of the parallel lines. COLOR=GRAYC0 signifies a grayc0 color is used for the background.

PATTERN7 VALUE=p2x30 COLOR=GRAY00 – p signifies a Pie chart slice, 2 is the shading level, x tells SAS to use cross-hatched lines, and 30 is the angle (degree) of the cross-hatched lines. COLOR=GRAY00 signifies a gray00 color is used for the background.

LIBNAME defines a library called mydata representing the folder of "c:\sas_exercise."

PROC GCHART DATA=MYDATA.as3 invokes SAS's GCHART function for the dataset named as3.sas7bdat in the mydata library.

PIE3D declares that the PIE3D chart procedure is to be used. The variable DOW is used to split the PIE for the variable vehicles (SUMVAR=vehicles). Actual data is the summation of Vehicles (TYPE=SUM). EXPLODE =" Monday" "Thursday" "Wednesday" tells SAS to separate these three slices out from the whole PIE.

NOHEADER tells SAS not to print the default title of "SUM of Vehicles by DOW."

SLICE=OUTSIDE instructs SAS to label slice outside.

VALUE=OUTSIDE tells SAS to label the slice with its data on the outside of the slice.

COUTLINE=GREY tells SAS to use gray-colored lines to demark the slice.

RUN instructs SAS to execute the PROC GCHART procedure.

QUIT terminates the procedure.

8

ODS, Title, and Footnotes

The material covered in this section is a basic introduction. The best way to gain an understanding of the fundamental practices associated with ODS, titles, and footnotes is by going through all the examples. Once you understand the principles behind these examples, you can further refine your output with more optional statements.

8.1 OUTPUT DELIVERY SYSTEM (ODS)

SAS procedures generate output, such as tables, graphics, and data. When we do not include any additional statements in the SAS Code File, the output takes the default format – HTML for SAS 9.3 and LISTING for earlier versions.

The Output Delivery System (ODS) offers you additional options to output your SAS procedure results in other formats such as LISTING (TXT), Rich Text Format (RTF), PostScript (PS), Portable Document Format (PDF), and the default Hypertext Markup Language (HTML).

When you explore ODS functions, make sure you understand the following basic ODS terms.

The first term is "Destination." With ODS, the Destination refers to both the file format (e.g., HTML, RTF, PDF, LISTING) and the actual file directory (e.g., c:\sas_exercise\my_sas_output). The second term is "Object." The Object refers to both the tabular data generated from a SAS procedure and the formatting instructions provided by the associated SAS procedure. The last term is "Style," focusing on visual effect and refers to font color, line thickness, table format, etc.

ODS does not analyze your data. It simply serves to output results in a different format and style.

Example: 8.1

Case A: Generating an Output with the PDF Format (The PDF Format has Additional Options You Can Use)

Case A: SAS Codes (E81a)

```
ODS PDF FILE='c:\sas_exercise\example8_default.pdf';
LIBNAME mydata 'c:\sas_exercise';
PROC MEANS DATA=mydata.speed8a;
VAR AMspeed PMspeed;
RUN;
ODS PDF CLOSE;
```

ODS invokes the Output Delivery System function by instructing SAS to deliver all outputs in a PDF file named example8_default.pdf to be stored in the folder of c:\sas_exercise.
LIBNAME defines a library named mydata representing the folder of c:\sas_exercise.
PROC MEANS DATA invokes the MEANS procedure for data speed8.sas7bdat stored in the mydata library.
VAR specifies the variables to be analyzed by the MEANS procedure are AMspeed and PMspeed.
RUN signals SAS to start the PROC MEANS procedure.
ODS PDF CLOSE terminates the ODS procedure and closes out the PDF output. ODS statements are always used in pairs.

Case A Result:

The SAS System

The MEANS Procedure

Variable	Label	N	Mean	Std Dev	Minimum	Maximum
AMspeed	AMspeed	41	36.4712195	12.0697045	16.5600000	56.9800000
PMspeed	PMspeed	41	55.6085366	14.3012347	22.3200000	75.6600000

Case B: PDF Options and NOPROCTITLE (Additional Options Are Used)
PDF Options include:

- Style= style name (e.g., PRINTER, Journal, Journal1…)
- STARTPAGE= Yes or No. When Yes, a page break is inserted between each procedure
- COLUMNS= n (number of columns for a columnar output).

```
ODS NOPROCTITLE;
```

Output does not mention (print out) SAS's PROC procedure name.

Case B: SAS Codes (E81b)

```
ODS PDF FILE='c:\sas_exercise\example8_NT_J.pdf' STYLE=Journal;
ODS NOPROCTITLE;
LIBNAME mydata 'c:\sas_exercise';
PROC MEANS DATA=mydata.speed8a;
VAR AMspeed PMspeed;
RUN;
ODS PDF CLOSE;
```

ODS invokes the Output Delivery System function by instructing SAS to deliver all outputs in a PDF file named example8_NT_J.pdf to be stored in the folder of c:\sas_exercise. The PDF file should have the style of Journal (Internal to SAS, different PDF styles are coded. A user's task is simply to tell SAS which format is to be used).

ODS NOPROCTITLE stands for ODS No Proc Title. It instructs SAS to stop providing the default "The MEANS Procedure" subtitle as a result of the PROC MEANS statement.

LIBNAME defines a library named mydata representing the folder of c:\sas_exercise.

PROC MEANS invokes the MEANS procedure for the speed8a.sas7bdat data stored in the mydata library.

VAR specifies the variables to be analyzed by the MEANS procedure are AMspeed and PMspeed.

RUN signals SAS to start the PROC MEANS procedure.

ODS PDF CLOSE terminates the ODS procedure and closes out the PDF output. ODS statements are always used in pairs.

Case B Result:

08:35 Tuesday, September 24, 2019 *1*

The SAS System

Variable	*Label*	*N*	*Mean*	*Std Dev*	*Minimum*	*Maximum*
AMspeed	AMspeed	20	49.6500000	16.5824098	24.0000000	74.0000000
PMspeed	PMspeed	20	58.8500000	15.2497627	24.0000000	78.0000000

Case C: Decimal Points Displayed and Multiple Destinations

Case C: SAS Codes (E81c)

```
ODS NOPROCTITLE;
ODS PDF FILE= 'c:\sas_exercise\example8_NT_J2.pdf' STYLE=Journal;
ODS LISTING FILE= 'c:\sas_exercise\example8_listing.txt';
LIBNAME mydata 'c:\sas_exercise';
PROC MEANS DATA=mydata.speed8a MAXDEC=2;
VAR AMspeed PMspeed;
RUN;
ODS LISTING CLOSE;
ODS PDF CLOSE;
```

ODS NOPROCTITLE stands for ODS NO Proc Title. It instructs SAS to stop providing the default "The MEANS Procedure" subtitle as a result of the PROC MEANS statement.

ODS PDF FILE invokes the Output Delivery System function by instructing SAS to deliver all outputs to a PDF FILE named example8_NT_J2.pdf to be stored in the folder of c:\sas_exercise. The PDF FILE should have the style of Journal (Internal to SAS, different PDF styles are coded. A user's task is simply to tell SAS which format is to be used).

ODS LISTING FILE invokes the Output Delivery System function by instructing SAS to deliver all outputs in a text file named example8_listing.txt to be stored in the folder of c:\sas_exercise.

LIBNAME defines a library named mydata representing the folder of c:\sas_exercise.

PROC MEANS DATA invokes the MEANS procedure for the speed8a.sas7bdat data stored in the mydata library.

VAR specifies the variables to be analyzed by the MEANS procedure are AMspeed and PMspeed.

RUN signals SAS to start the PROC MEANS procedure.

ODS LISTING CLOSE terminates the ODS procedure and closes out the LISTING output. It is always used in pairs with the initial ODS statement.

ODS PDF CLOSE terminates the ODS procedure and closes out the PDF output. ODS statements are always used in pairs.

Case C Result:

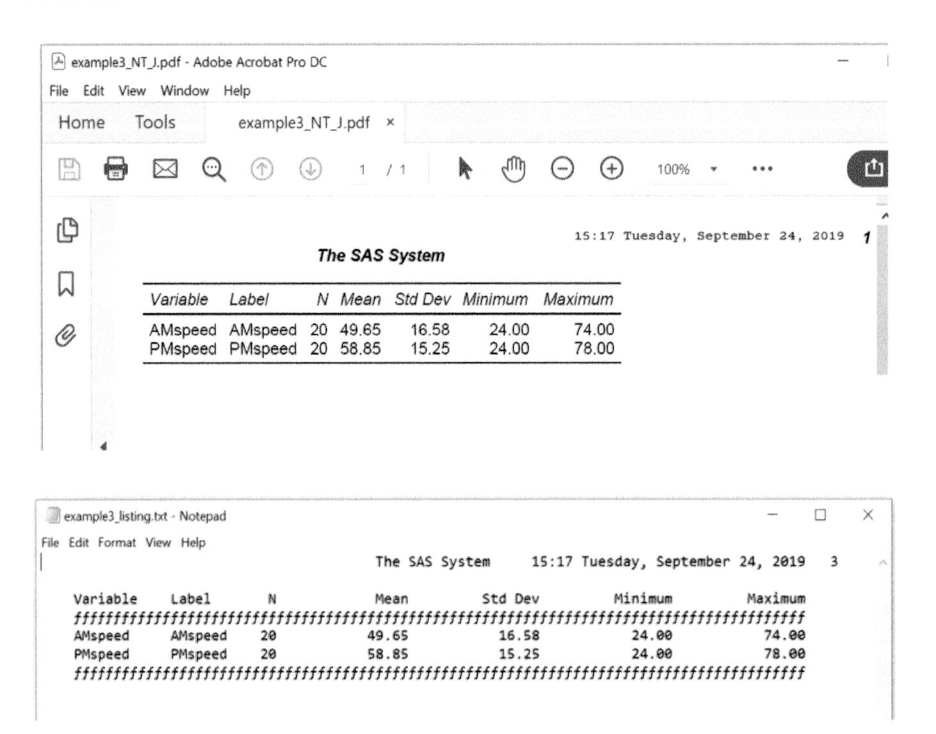

example3_NT_J.pdf - Adobe Acrobat Pro DC

15:17 Tuesday, September 24, 2019 1

The SAS System

Variable	Label	N	Mean	Std Dev	Minimum	Maximum
AMspeed	AMspeed	20	49.65	16.58	24.00	74.00
PMspeed	PMspeed	20	58.85	15.25	24.00	78.00

example3_listing.txt - Notepad

```
                                   The SAS System     15:17 Tuesday, September 24, 2019   3

 Variable     Label        N              Mean          Std Dev          Minimum          Maximum
 ƒƒƒƒƒƒƒƒƒƒƒƒƒƒƒƒƒƒƒƒƒƒƒƒƒƒƒƒƒƒƒƒƒƒƒƒƒƒƒƒƒƒƒƒƒƒƒƒƒƒƒƒƒƒƒƒƒƒƒƒƒƒƒƒƒƒƒƒƒƒƒƒƒƒƒƒƒƒƒƒƒƒƒƒƒƒƒƒƒƒƒƒ
 AMspeed      AMspeed     20             49.65            16.58            24.00            74.00
 PMspeed      PMspeed     20             58.85            15.25            24.00            78.00
 ƒƒƒƒƒƒƒƒƒƒƒƒƒƒƒƒƒƒƒƒƒƒƒƒƒƒƒƒƒƒƒƒƒƒƒƒƒƒƒƒƒƒƒƒƒƒƒƒƒƒƒƒƒƒƒƒƒƒƒƒƒƒƒƒƒƒƒƒƒƒƒƒƒƒƒƒƒƒƒƒƒƒƒƒƒƒƒƒƒƒƒƒ
```

Case D: "Excel" Data Output (Where the File Can Be Opened by Excel)

Case D: SAS Codes (E81d)

```
ODS HTML FILE='c:\sas_exercise\example_c8.xls';
ODS NOPROCTITLE;
libname my 'c:\sas_exercise';
PROC PRINT DATA=my.speed8a;
RUN;
ODS HTML CLOSE;
```

ODS HTML FILE invokes the Output Delivery System function to establish a file named example_c8.xls.

ODS NOPROCTITLE stands for ODS No Proc Title. It instructs SAS to stop providing the default "The Print Procedure" subtitle as a result of the PROC PRINT DATA statement.

libname defines a library named my representing the folder of c:\sas_exercise.

PROC PRINT DATA invokes the PRINT procedure for the speed8.sas7bdat data stored in the my library.

RUN signals SAS to start the PROC PRINT procedure.

ODS HTML CLOSE terminates the ODS procedure. ODS statements are always used in pairs.

Case D Result:

The SAS System

Obs	Location	RoadName	Direction	RoadType	MeasureDate	AMspeed	PMspeed
1	PG569	I95	N	1	1-Jan-19	24.15	43.65
2	PG569	I95	N	1	2-Jan-19	28.98	68.87
3	PG569	I95	N	1	3-Jan-19	22.08	67.9
4	PG569	I95	N	1	4-Jan-19	40.02	53.35
5	PG569	I95	N	1	5-Jan-19	49.68	75.66
6	PG569	I95	N	1	6-Jan-19	51.06	69.84
7	PG569	I95	N	1	7-Jan-19	41.4	62.08
8	PG569	I95	N	1	8-Jan-19	24.84	60.14
9	PG569	I95	N	1	9-Jan-19	31.05	60.14
10	PG569	I95	N	1	10-Jan-19	19.32	31.04
11	PG569	I95	N	1	11-Jan-19	24.84	44.62
12	PG569	I95	N	1	12-Jan-19	48.3	65.96
13	PG569	I95	N	1	13-Jan-19	48.99	70.81
14	PG569	I95	N	1	14-Jan-19	38.64	58.2
15	PG569	I95	N	1	15-Jan-19	31.74	58.2
16	PG569	I95	N	1	16-Jan-19	26.22	40.74
17	PG569	I95	N	1	17-Jan-19	24.84	23.28
18	PG569	I95	N	1	18-Jan-19	16.56	43.65
19	PG569	I95	N	1	20-Jan-19	45.54	75.66
20	PG569	I95	S	1	21-Jan-19	46.92	67.9
21	PG569	I95	S	1	21-Jan-19	45.54	43.65
22	PG569	I95	S	1	1-Jan-19	26.95	41.85
23	PG569	I95	S	1	2-Jan-19	32.34	66.03

Case E: Multiple PROC Procedures with the PDF Journal Option

Case E: SAS Codes (E81e)

```
ODS PDF FILE='c:\sas_exercise\example8_J.pdf' STYLE=Journal;
ODS NOPROCTITLE;
```

```
LIBNAME mydata 'c:\sas_exercise';

PROC MEANS DATA=mydata.speed8a;
VAR AMspeed PMspeed;
RUN;

PROC Univariate Data=mydata.speed8a;
VAR AMspeed;
RUN;
ODS PDF CLOSE;
```

ODS PDF FILE invokes the Output Delivery System function by instructing SAS to deliver all outputs in a PDF file named example8_J.pdf to be stored in the folder of c:\sas_exercise. The PDF file should have the style of Journal (Internal to SAS, different PDF styles are coded. A user's task is simply to tell SAS which format is to be used).

ODS NOPROCTITLE stands for ODS NO Proc Title. It instructs SAS to stop providing the default "The MEANS Procedure" and the "The Univariate Procedure" subtitles as a result of the PROC statements later.

LIBNAME defines a library named mydata representing the folder of c:\sas_exercise.

PROC MEANS DATA invokes the MEANS procedure for the speed8.sas7bdat data stored in the mydata library.

VAR specifies the variables to be analyzed by the MEANS procedure are AMspeed and PMspeed.

RUN signals SAS to start the PROC MEANS procedure.

PROC Univariate Data invokes the Univariate procedure for the speed8a.sas7bdat data stored in the mydata library.

VAR specifies the variable to be analyzed by the Univariate procedure is the AMspeed.

RUN signals SAS to start the PROC UNIVARIATE procedure.

ODS PDF CLOSE terminates the ODS procedure. ODS statements are always used in pairs.

Case E Result:

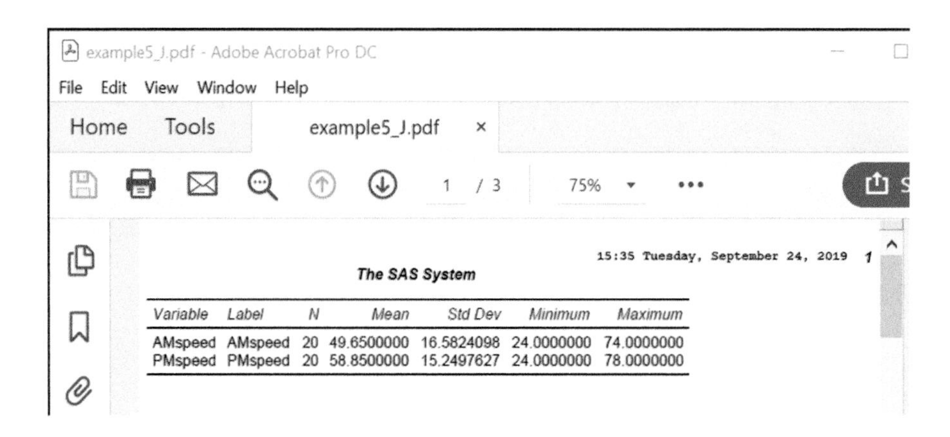

Variable	Label	N	Mean	Std Dev	Minimum	Maximum
AMspeed	AMspeed	20	49.6500000	16.5824098	24.0000000	74.0000000
PMspeed	PMspeed	20	58.8500000	15.2497627	24.0000000	78.0000000

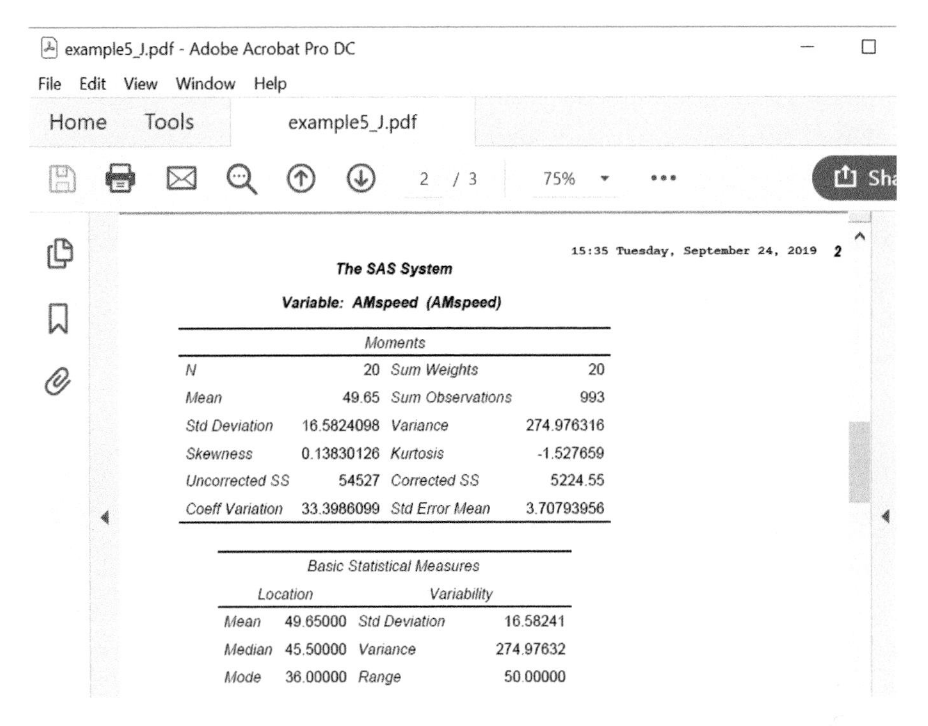

15:35 Tuesday, September 24, 2019 2

The SAS System

Variable: AMspeed (AMspeed)

Moments			
N	20	Sum Weights	20
Mean	49.65	Sum Observations	993
Std Deviation	16.5824098	Variance	274.976316
Skewness	0.13830126	Kurtosis	-1.527659
Uncorrected SS	54527	Corrected SS	5224.55
Coeff Variation	33.3986099	Std Error Mean	3.70793956

Basic Statistical Measures			
Location		Variability	
Mean	49.65000	Std Deviation	16.58241
Median	45.50000	Variance	274.97632
Mode	36.00000	Range	50.00000

Case F: "ODS Trace On" Statement

The "ODS Trace On" statement provides additional information in the SAS log file regarding output data, tables, and variables from the executed SAS procedure by creating a record of each output Object that has been created.

Case F1: No "ODS TRACE ON" Statement – default

Case F1: SAS Codes (E81f1)

```
LIBNAME mydata 'c:\
sas_exercise';
PROC SORT DATA=mydata.speed8a;
BY direction;
PROC MEANS DATA=mydata.speed8a;
VAR AMspeed PMspeed;
BY direction;
RUN;
```

LOGFILE

```
SAS/STAT 12.1, SAS/ETS 12.1

NOTE: SAS initialization used:
      real time           1.22
seconds
      cpu time            1.04
seconds
```

LIBNAME defines a library named mydata representing the folder of c:\sas_exercise.

PROC SORT DATA tells SAS to sort the data named speed8a.sas7bdat in the mydata library BY direction

PROC MEANS invokes the MEANS procedure for the speed8a.sas7bdat data stored in the mydata library.

VAR specifies the variables to be analyzed by the MEANS procedure are AMspeed and PMspeed.

BY direction tells SAS to compute the MEANS by direction.

RUN signals SAS to start the PROC MEANS procedure.

The execution of the above codes results in a standard SAS LOGFILE where additional information covering the creation of other files is not present.

```
1     LIBNAME mydata 'c:\sas_exercise';
NOTE: Libref MYDATA was successfully assigned as follows:
      Engine:        V9
      Physical Name: c:\sas_exercise
2     PROC SORT DATA=mydata.speed8a;
3     BY direction;

NOTE: Input data set is already sorted, no sorting done.
NOTE: PROCEDURE SORT used (Total process time):
      real time           0.01 seconds
      cpu time            0.01 seconds


4     PROC MEANS DATA=mydata.speed8a;
5     VAR AMspeed PMspeed;
6     BY direction;
7     RUN;

NOTE: Writing HTML Body file: sashtml.htm
NOTE: There were 41 observations read from the data set MYDATA.
SPEED8A.
NOTE: PROCEDURE MEANS used (Total process time):
      real time           0.38 seconds
      cpu time            0.12 seconds
```

Case F1 Result:

The MEANS Procedure

Direction=N

Variable	Label	N	Mean	Std Dev	Minimum	Maximum
AMspeed	AMspeed	19	33.5921053	11.3496381	16.5600000	51.0600000
PMspeed	PMspeed	19	56.5152632	14.9708644	23.2800000	75.6600000

Direction=S

Variable	Label	N	Mean	Std Dev	Minimum	Maximum
AMspeed	AMspeed	22	38.9577273	12.3730448	18.4800000	56.9800000
PMspeed	PMspeed	22	54.8254545	14.0027527	22.3200000	72.5400000

Case F2: "ODS TRACE ON" Is Used

Case F2: SAS Codes (E81f2)

```
ODS TRACE ON;
LIBNAME mydata 'c:\sas_exercise';
PROC SORT DATA=mydata.speed8a;
BY direction;
PROC MEANS DATA=mydata.speed8a;
VAR AMspeed PMspeed;
BY direction;
RUN;
ODS TRACE OFF;
```

ODS TRACE ON instructs SAS to keep the full log while executing all the statements.

LIBNAME defines a library named mydata representing the folder of c:\sas_exercise.

PROC SORT DATA tells SAS to sort the data named speed8a.sas7bdat in the mydata library by direction.

PROC MEANS DATA invokes the MEANS procedure for the speed8a.sas7bdat data stored in the mydata library.

VAR specifies the variables to be analyzed by the MEANS procedure are AMspeed and PMspeed.

By direction tells SAS to compute the MEANS by direction.

RUN signals SAS to start the PROC MEANS procedure.

ODS TRACE OFF instructs SAS to terminate the TRACE process.

The execution of the above codes results in a standard SAS LOGFILE and additional information covering the creation of all output files.

Case F2 Result:

The MEANS Procedure

Direction=N

Variable	Label	N	Mean	Std Dev	Minimum	Maximum
AMspeed	AMspeed	19	33.5921053	11.3496381	16.5600000	51.0600000
PMspeed	PMspeed	19	56.5152632	14.9708644	23.2800000	75.6600000

Direction=S

Variable	Label	N	Mean	Std Dev	Minimum	Maximum
AMspeed	AMspeed	22	38.9577273	12.3730448	18.4800000	56.9800000
PMspeed	PMspeed	22	54.8254545	14.0027527	22.3200000	72.5400000

LOGFILE

```
SAS/STAT 12.1, SAS/ETS 12.1

NOTE: SAS initialization used:
      real time           1.37 seconds
      cpu time            1.17 seconds
```

```
1     ODS TRACE ON;
2     LIBNAME mydata 'c:\sas_exercise';
NOTE: Libref MYDATA was successfully assigned as follows:
      Engine:        V9
      Physical Name: c:\sas_exercise
3     PROC SORT DATA=mydata.speed8a;
4     BY direction;

NOTE: Input data set is already sorted, no sorting done.
NOTE: PROCEDURE SORT used (Total process time):
      real time           0.02 seconds
      cpu time            0.01 seconds


5     PROC MEANS DATA=mydata.speed8a;
6     VAR AMspeed PMspeed;
7     BY direction;
8     RUN;

NOTE: Writing HTML Body file: sashtml.htm

Output Added:
-------------
Name:       Summary
Label:      Summary statistics
Template:   base.summary
Path:       Means.ByGroup1.Summary
-------------
NOTE: The above message was for the following BY group:
      Direction=N

Output Added:
-------------
Name:       Summary
Label:      Summary statistics
Template:   base.summary
Path:       Means.ByGroup2.Summary
-------------
NOTE: The above message was for the following BY group:
      Direction=S
NOTE: There were 41 observations read from the data set MYDATA.
SPEED8A.
NOTE: PROCEDURE MEANS used (Total process time):
      real time           0.48 seconds
      cpu time            0.15 seconds

9     ODS TRACE OFF
```

Case G: Using "Select" to Obtain a Specific Output

In this example, I am using information contained in the full logfile as a result of using the "ODS Trace on" command to carryout additional task otherwise would not be able to accomplish.

Case G: SAS Codes (E81g)

```
ODS TRACE ON;
LIBNAME mydata 'c:\sas_exercise';
PROC SORT DATA=mydata.speed8a;
BY direction;
PROC MEANS DATA=mydata.speed8a;
VAR AMspeed PMspeed;
BY direction;
ODS SELECT Means.ByGroup2.Summary;
RUN;
ODS TRACE OFF;
```

ODS TRACE ON instructs SAS to keep the full log while executing all the statements.

LIBNAME defines a library named mydata representing the folder of c:\sas_exercise.

PROC SORT DATA tells SAS to SORT the data named speed8a.sas7bdat in the mydata library by direction.

PROC MEANS DATA invokes the MEANS procedure for the speed8a.sas7bdat data stored in the mydata library.

VAR specifies the variables to be analyzed by the MEANS procedure are AMspeed and PMspeed.

BY direction tells SAS to compute the MEANS by direction.

RUN signals SAS to start the PROC MEANS procedure.

ODS SELECT directs SAS to output only the desired output Means.ByGroup2.Summary. The Means/ByGroup2.Summary information is shown in the full log file. Without seeing it in the full log file, we would not now its existence.

ODS TRACE OFF instructs SAS to terminate the TRACE ON process.

The execution of the above codes results in a standard SAS LOGFILE and additional information covering the creation of all output files.

Case G Result:

The MEANS Procedure

Direction=S

Variable	Label	N	Mean	Std Dev	Minimum	Maximum
AMspeed	AMspeed	22	38.9577273	12.3730448	18.4800000	56.9800000
PMspeed	PMspeed	22	54.8254545	14.0027527	22.3200000	72.5400000

8.2 TITLE AND FOOTNOTE

Title and footnote help to identify results and outputs. You can have up to a maximum of 10 titles and 10 footnotes in a single SAS run. The statements to add a title or footnote to your output follow the general format as illustrated below:

TITLE<number> <text-options> <"text"> ;
FOOTNOTE<n> <text-options> <"text">;

Where

- number specifies the relative line that contains the title/footnote
- text-options specifies formatting options for the text – color, font, size, etc
- "text" specifies text string that can contain up to 512 characters.

You can use title or footnote statements anywhere in your SAS code. Both are global statements.

Case A: Only One Title and One Footnote with Height Options

Case A: SAS Codes (E82a)

```
TITLE HEIGHT=16pt 'Result Generated By the PROC Means and ODS
Select' ;
FOOTNOTE HEIGHT=12pt 'Prepared by ABC at Lab1' ;
ODS NOPROCTITLE;
ODS TRACE ON;
LIBNAME mydata 'c:\sas_exercise';
PROC SORT DATA=mydata.speed8a;
BY direction;
PROC MEANS DATA=mydata.speed8a;
VAR AMspeed PMspeed;
BY direction;
ODS SELECT Means.ByGroup2.Summary;
RUN;
ODS TRACE OFF;
```

> *TITLE invokes the SAS TITLE function. HEIGHT defines the font size. The actual title is "Result Generated By the PROC Means and ODS Select"*
>
> *FOOTNOTE invokes the SAS FOOTNOTE function. HEIGHT defines the font size. The actual title is "Prepared by ABC at Lab1"*
>
> *ODS NOPROCTITLE stands for ODS No Proc Title and instructs SAS not to generate the default PROC Procedure title (e.g., "The Means Procedure" under PROC MEANS).*
>
> *ODS TRACE ON instructs SAS to keep the full log while executing all the statements.*
>
> *LIBNAME defines a library named mydata representing the folder of c:\sas_exercise.*

PROC SORT DATA tells SAS to sort the data named speed8a.sas7bdat in the mydata library by direction.

PROC MEANS DATA invokes the MEANS procedure for the speed8a.sas7bdat data stored in the mydata library.

VAR specifies the variables to be analyzed by the MEANS procedure are AMspeed and PMspeed.

By direction tells SAS to compute the MEANS by direction.

RUN signals SAS to start the PROC MEANS procedure.

ODS SELECT directs SAS to output only the desired output Means.ByGroup2.Summary (see LOGFILE).

ODS TRACE OFF instructs SAS to terminate the TRACE process.

Case A Result:

Result Generated By the PROC Means and ODS Select

Direction=S

Variable	Label	N	Mean	Std Dev	Minimum	Maximum
AMspeed	AMspeed	22	38.9577273	12.3730448	18.4800000	56.9800000
PMspeed	PMspeed	22	54.8254545	14.0027527	22.3200000	72.5400000

Prepared by ABC at Lab1

Case B: Two Titles and Two Footnotes without Numbering

When you specify multiple titles or footnotes without uniquely numbering them, the last title overrides all previous titles, and the last footnote overrides all previous footnotes.

Case B: SAS Codes (E82b)

```
TITLE HEIGHT=16pt 'Result Generated By the PROC Means and ODS
Select' ;
TITLE HEIGHT=14PT 'SECOND TITLE';
FOOTNOTE HEIGHT=12pt 'Prepared by ABC at Lab1' ;
FOOTNOTE HEIGHT=12PT 'SECOND FOOTNOTE';
ODS NOPROCTITLE;
ODS TRACE ON;
LIBNAME mydata 'c:\sas_exercise';
PROC SORT DATA=mydata.speed8a;
BY direction;
PROC MEANS DATA=mydata.speed8a;
VAR AMspeed PMspeed;
BY direction;
ODS SELECT Means.ByGroup2.Summary;
RUN;
ODS_TRACE OFF;
```

TITLE invokes the SAS TITLE function. HEIGHT defines the font size to be 16 points. The actual title is " Result Generated By the PROC Means and ODS Select "

TITLE invokes the SAS TITLE function. HEIGHT defines the font size to be 14 points. The actual title is "SECOND TITLE." This TITLE statement overrides the prior one given they are presented under the identical TITLE command.

FOOTNOTE invokes the SAS FOOTNOTE function. HEIGHT defines a 12-point font size. The actual title is "Prepared by ABC at Lab1"

FOOTNOTE invokes the SAS FOOTNOTE function. HEIGHT defines a 12-point font size. The actual title is " SECOND FOOTNOTE." This FOOTNOTE overrides the previous one given both of the footnotes are under the same FOOTNOTE command.

ODS TRACE ON instructs SAS to keep the full log while executing all the statements.

ODS NOPROCTITLE stands for ODS NO Proc Title and instructs SAS not to generate the default PROC Procedure title (e.g., "The Means Procedure" under PROC MEANS).

LIBNAME defines a library named mydata representing the folder of c:\sas_exercise.

PROC SORT DATA tells SAS to sort the data named speed8a.sas7bdat in the mydata library BY direction.

PROC MEANS DATA invokes the MEANS procedure for the speed8a.sas7bdat data stored in the mydata library.

VAR specifies the variables to be analyzed by the MEANS procedure are AMspeed and PMspeed.

BY direction tells SAS to compute the MEANS based on direction.

RUN signals SAS to start the PROC MEANS procedure.

ODS TRACE OFF instructs SAS to terminate the TRACE ON process.

Case B Result:

SECOND TITLE

Direction=S

Variable	Label	N	Mean	Std Dev	Minimum	Maximum
AMspeed	AMspeed	22	38.9577273	12.3730448	18.4800000	56.9800000
PMspeed	PMspeed	22	54.8254545	14.0027527	22.3200000	72.5400000

SECOND FOOTNOTE

Case C: Three Titles and Two Footnotes with Unique Sequential Numberings

Case C: SAS Codes (E82c)

```
TITLE HEIGHT=16pt 'Result Generated By the PROC Means and ODS
Select' ;
TITLE2 HEIGHT=12PT 'MY SECOND TITLE';
FOOTNOTE HEIGHT=12pt 'Prepared by ABC at Lab1' ;
ODS NOPROCTITLE;
ODS TRACE ON;
LIBNAME mydata 'c:\sas_exercise';
```

```
TITLE3 Height=14 'My 3rd Title';
PROC SORT DATA=mydata.speed8a;
BY direction;
FOOTNOTE2 HEIGHT=10PT '2ND FOOTNOTE';
PROC MEANS DATA=mydata.speed8a;
VAR AMspeed PMspeed;
BY direction;
ODS SELECT Means.ByGroup2.Summary;
RUN;
ODS TRACE OFF;
```

TITLE invokes the SAS TITLE function. HEIGHT defines the font size to be 16 points. The actual title is " Result Generated By the PROC Means and ODS Select"

TITLE2 invokes the SAS TITLE function. HEIGHT defines the font size to be 12 points. The actual title is "MY SECOND TITLE."

FOOTNOTE invokes the SAS FOOTNOTE function. HEIGHT defines a 12-point font size. The actual title is " Prepared by ABC at Lab1"

ODS TRACE ON instructs SAS to keep the full log while executing all the statements.

ODS NOPROCTITLE stands for ODS No Proc Title and instructs SAS not to generate the default PROC Procedure title (e.g., "The Means Procedure" under PROC MEANS).

LIBNAME defines a library named mydata representing the folder of c:\sas_exercise.

TITLE3 invokes the SAS TITLE function. HEIGHT defines the font size to be 14 points. The actual title is "My 3rd Title." The order of where a TITLE statement is declared does not matter for a SAS procedure. By default, SAS allows a total of ten titles.

PROC SORT DATA tells SAS to sort the data named speed8a.sas7bdat in the mydata library by direction.

FOOTNOTE2 invokes the SAS FOOTNOTE function. HEIGHT defines a 10-point font size. The actual title is "2ND FOOTNOTE." The order of where a FOOTNOTE statement is declared does not matter for a SAS procedure. By default, SAS allows ten footnotes.

PROC MEANS DATA invokes the MEANS procedure for the speed8a.sas7bdat data stored in the mydata library.

VAR specifies the variables to be analyzed by the MEANS procedure are AMspeed and PMspeed.

BY direction tells SAS to compute the MEANS based on direction.

RUN signals SAS to start the PROC MEANS procedure.

ODS TRACE OFF instructs SAS to terminate the TRACE ON process.

Case C Result:

Result Generated By the PROC Means and ODS Select
MY SECOND TITLE

My 3rd Title

Direction=S

Variable	Label	N	Mean	Std Dev	Minimum	Maximum
AMspeed	AMspeed	22	38.9577273	12.3730448	18.4800000	56.9800000
PMspeed	PMspeed	22	54.8254545	14.0027527	22.3200000	72.5400000

Prepared by ABC at Lab1
2ND FOOTNOTE

Footnotes and titles make your report more complete and easier to read and understand. In addition to the font sizes (HEIGHT) illustrated in the above example, you can also specify font type and color.

9

Behind the Scenes Logic on Data Reading

A good understanding of how SAS reads your data will help you code your program more efficiently.

When SAS reads your data in the DATA step, there are two steps involved. The first is the compilation, and the second is the execution.

During compilation, SAS scans all statements contained in the DATA procedure for syntactical/grammatical errors. During this process, SAS also establishes the Program Data Vector (PDV) (a set of memory on your computer) and additional descriptor information as listed below.

A: Two temporary automatic variables called _N_ and _ERROR_. The _N_ represents the number of observations SAS will be processing when the process is in its execution stage. The _ERROR_ represents whether an error has occurred during the processing of observation and the execution of the code.

While _N_ ranges from 1 to whatever the number of observations your data has, the _ERROR_ has a binary choice of 0 or 1. When an error is encountered, it is assigned a 1. Otherwise, it is assigned a 0.

B: One space for each of the variables listed

C: One space for each of the variables created in the DATA step.

See the example below for further clarification.

Example 9: Original Input Data

vehicle.csv:

```
Book2.csv - Notepad
File  Edit  Format  View  Help
region,population,truck,cars,total ,rate
Waye,425678,4678,255401,260079,0.61
Clemson,253652,2213,153121,155334,0.61
Apple,1269,98,698,796,0.63
Beaver,34562,3986,19892,23878,0.69
Leon,1239,121,1028,1149,0.92
```

SAS Codes – illustrating how SAS reads data

```
DATA vehicles
INFILE 'c:\sas_exercise\vehicle.csv' DLM=',' FIRSTOBS=2;
INPUT region $ population truck cars;
Total=truck+cars;
Rate=total/population;
RUN;
```

With the above DATA procedure, SAS creates a SAS dataset called "vehicles" by reading from an external vehicle.csv data file located in the folder of c:\sas_exercise. The data has four variables: region, population, truck, and cars. During the compilation stage, SAS creates two new variables: "Total" and "Rate" as instructed by the assignment statements in the DATA step. SAS also creates two additional temporary variables _ERROR_ and _N_.

The compilation process first checks the correctness of all statements contained between the DATA and RUN statements, which are as follows:

```
INFILE 'c:\sas_exercise\vehicle.csv' DLM=',' FIRSTOBS=2;
INPUT region $ population truck cars;
Total=truck+cars;
Rate=total/population;
```

Once the syntax check is done, SAS creates a PDV, as illustrated below.

_N_	_ERROR_	region	population	truck	cars	total	rate

During the execution stage, the DATA step statements are executed one observation (one row) at a time. The actual observation is read from the original file to a buffer and then to the PDV. From the PDV, a dataset is created, as illustrated below.

When the first observation row is read, the PDV appears as:

_N_	_ERROR_	region	population	truck	cars	total	rate
1	0	Waye	425678	4678	255401	260079	0.61

And the eventual SAS data is:

region	population	truck	cars	total	rate
Waye	425678	4678	255401	260079	0.61

When SAS reads the second row, the PDV appears as:

_N_	_ERROR_	region	population	truck	cars	total	rate
1	0	Waye	425678	4678	255401	260079	0.61
2	0	Clemson	253652	2213	153121	155334	0.61

And the SAS data now is:

region	population	truck	cars	total	rate
Waye	425678	4678	255401	260079	0.61
Clemson	253652	2213	153121	155334	0.61

SAS continues the row-by-row reading until all rows (observations) are read, as illustrated below.

_N_	_ERROR_	region	population	truck	cars	total	rate
1	0	Waye	425678	4678	255401	260079	0.61
2	0	Clemson	253652	2213	153121	155334	0.61
3	0	Apple	1269	98	698	796	0.63
6	0	Beaver	34562	3986	19892	23878	0.69
7	0	Leon	1239	121	1028	1149	0.92
686	0	Polk	27893	3012	2094	5106	0.18

Final SAS Dataset:

region	population	truck	cars	total	rate
Waye	425678	4678	255401	260079	0.61
Clemson	253652	2213	153121	155334	0.61
Apple	1269	98	698	796	0.63
Beaver	34562	3986	19892	23878	0.69
Leon	1239	121	1028	1149	0.92
....					

Keep in mind that all statements associated with the DATA procedure are executed row by row (observations by observations). Only when all statements are completed will SAS then move to process the next row.

Index